PHOTOCATALYTIC PRODUCTION OF ENERGY-RICH COMPOUNDS

 Third EC Biomass Development Programme
Series editor: *W. Palz*

ENERGY FROM BIOMASS 2

Proceedings of Second EC Workshop on Photochemical and Photobiological Processes for the Production of Energy-Rich Compounds, Seville, Spain, 22–25 September, 1987

PHOTOCATALYTIC PRODUCTION OF ENERGY-RICH COMPOUNDS

Edited by

G. GRASSI

Commission of the European Communities, Brussels, Belgium

and

D. O. HALL

King's College, London, UK

ELSEVIER APPLIED SCIENCE
LONDON and NEW YORK

ELSEVIER APPLIED SCIENCE PUBLISHERS LTD
Crown House, Linton Road, Barking, Essex IG11 8JU, England

Sole Distributor in the USA and Canada
ELSEVIER SCIENCE PUBLISHING CO., INC.
52 Vanderbilt Avenue, New York, NY 10017, USA

WITH 19 TABLES AND 99 ILLUSTRATIONS

© 1988 ECSC, EEC, EAEC, BRUSSELS AND LUXEMBOURG

British Library Cataloguing in Publication Data

E.C. Workshop on Photochemical and Photobiological
Processes for the Production of Energy-rich
Compounds; 2nd (1987), Seville, Spain.
Photocatalytic production of energy-rich
compounds.
1. Alternative energy sources. Photochemical
conversion
I. Title II. Hall, D.O. (David Oakley)
III. Grassi, G., *1929–*
621.042

ISBN 1-85166-216-2

Library of Congress CIP data applied for

Publication arrangements by Commission of the European Communities, Directorate-General Telecommunications, Information Industries and Innovation, Luxembourg.

EUR 11371

LEGAL NOTICE

Neither the Commission of the European Communities nor any person acting on behalf of the Commission is responsible for the use which might be made of the following information.

No responsibility is assumed by the Publisher for any injury and/or damage to persons or property as a matter of products liability, negligence or otherwise, or from any use or operation of any methods, products, instructions or ideas contained in the material herein.

Special regulations for readers in the USA

This publication has been registered with the Copyright Clearance Center Inc. (CCC), Salem, Massachusetts. Information can be obtained from the CCC about conditions under which photocopies of parts of this publication may be made in the USA. All other copyright questions, including photocopying outside the USA, should be referred to the publisher.

All rights reserved. No part of this publication may be reproduced, stored in a retrieval system, or transmitted in any form or by any means, electronic, mechanical, photocopying, recording, or otherwise, without the prior written permission of the publisher.

Printed in Great Britain by Galliard (Printers) Ltd, Great Yarmouth

Preface

This workshop comprises part of the four-year (1985–1988) non-nuclear energy R & D programme for the development of renewable energy sources which is being implemented by the Commission of the European Communities (Directorate-General Science, Research and Development—DG XII). The aim of the workshop was to present work by the contracting laboratories in addition to work by numerous other research laboratories in 11 European countries. Extensive discussions were also held on the present status of this basic, directed research in photochemistry, photoelectrochemistry and photobiology and where the future emphasis may usefully lie.

Thus the book presents the proceedings of all the papers presented and summarizes the recommendations made by the participants as to where future research support may be most effectively placed. It was emphasized in these recommendations that the interdisciplinary collaboration between photochemistry and photobiology had been quite successfully achieved in this European programme. There were both high quality basic research and practical benefits accruing from the work—these are elucidated in the report on proposed areas for future research.

We hope that this collaborative programme will continue in bringing together photochemists and photobiologists to help to ultimately devise practical systems for solar energy conversion and storage of useful compounds for energy and chemicals.

G. Grassi
Commission of the European Communities

D. O. Hall
King's College, London

2nd EEC Workshop on Photochemical and Photobiological Processes
for Producing Energy Rich Compounds
Seville, Spain; September 22nd-25th, 1987

*Recommendations from the participants for future research
in the R & D programme, DGXII/F-4*

The main areas proposed for further research support are:

1. Emphasise work on reaction centres and light-harvesting mechanisms, chemical and biological. Studies of light-activated reactions on surfaces and at interfaces. Synthesis and study of model compounds which mimic photosynthetic reaction centres, light-harvesting complexes and associated enzymes - this may provide very interesting new materials for further development.

2. The research programme should parallel and overlap research in photovoltaics and semiconductors, and thus provide a unique link and basis of information across the fields of photovoltaics, semiconductors and photosynthesis.

3. Research on multi-electron transfer reactions. These are different mechanisms from photovoltaics where single electron transfer reactions occur.

4. Activation and conversion of small molecules; photocatalysis. Production of H_2 from protons, N_2 fixation for ammonia production, CO_2 fixation for organics, sulphide (H_2S and Na_2S) oxidation, methane activation for methanol production, etc.

5. Photochemical processes for solving pollution problems. Removal of organic pollutants and inorganic molecules (e.g. NO_3 and NO_2) from water.

6. Genetic manipulation of cyanobacteria. Genetic engineering techniques with these micro-organisms are rapidly advancing and will make the understanding of photosynthetic reaction centre and water-splitting mechanisms easier, especially from the recent work in the USA and in Japan. However, there is no significant European group working in this field - this should be rectified.

7. Photobiotechnology. Photosynthetic micro-organisms specifically designed for the production of compounds such as carotenoids, ammonia, amino-acids, H_2 and possibly ethanol. Cyanobacteria, for example, are versatile micro-organisms whose growth and metabolism can be manipulated to preferentially excrete biochemicals.

8. Prevention of photodestruction. Many industrial products (e.g. paints and plastics) are destroyed by continuous exposure to the sun. Semiconductors and photoelectrodes are corroded in continuous light. Biological materials are also prone to photoinhibition. Understanding the mechanisms of photodecay (corrosion and inhibition) is essential to prevent such wasteful processes in chemical and biological systems.

Contents

Preface v

I. PRESENTATIONS BY EC CONTRACTORS

Stabilisation of Si photoanodes by silicon oxynitride coatings: attempted synthesis of binuclear Co complexes for (photo)electrochemical reduction of CO_2 1
    A. Mackor, F. Verbeek, T. P. M. Koster, C. I. M. A. Spee and C. W. de Kreuk

Studies on isolated plant pigment–protein complexes 10
    G. Porter and L. Giorgi

Primary reactions in plant photosynthetic reaction centers . . . 20
    Paul Mathis

Immobilized photosynthetic systems for the production of ammonia and photocurrents 28
    D. O. Hall, K. K. Rao, H. de Jong, M. Gratzel and M. C. W. Evans

II. PHOTOCHEMISTRY

Towards the design of molecular photochemical devices based on ruthenium bipyridine photosensitizer units 41
    Franco Scandola

Photoinduced charge-separation in models for photosynthesis . . 51
    J. W. Verhoeven, H. Oevering, M. N. Paddon-Row, J. Kroon and A. G. M. Kunst

Electron transfer photosensitized by zinc porphyrins in reversed micelles   59
    Silvia M. B. Costa

Efficient visible light sensitization of TiO_2 by surface complexation with transition metal cyanides 69
    E. Vrachnou, N. Vlachopoulos and M. Grätzel

Photo-induced electron transfer reactions in polymer-bound ruthenium bipyridyl complexes 80
Patricia M. Ennis and John M. Kelly

Self organization and photofunctionalization of supramolecular systems: photosensitive polymeric monolayers, multilayers and liposomes . . 85
L. Häussling, M. Haubs, H. Ringsdorf and J. Schneider

Photogeneration of hydrogen: the photochemical way of storing solar energy 97
G. Munuera, A. Fernández and J. P. Espinós

Inorganic photosynthesis: the photofixation of the atmospheric dinitrogen on transition metal oxides 105
R. I. Bickley and J. A. Navio-Santos

Recent trends in the search for new photosensitizers 115
A. Juris and V. Balzani

Reactivity of CO_2 and related heterocumulenes towards transition metal compounds 122
E. Carmona, A. Galindo and M. A. Muñoz

III. PHOTOELECTROCHEMISTRY

Medium effects upon the stability of n-GaAs-based photoelectrochemical cells 127
S. Lingier, D. Vanmaekelbergh and W. P. Gomes

Unpinning of energy bands in photoelectrochemical cells: a consequence of surface chemistry and surface charge 138
D. Meissner and R. Memming

H_2O_2 production in a photoelectrochemical cell with TiO_2 electrodes: reaction mechanisms and efficiency 148
D. Tafalla and P. Salvador

Adsorption experiments for modelling semiconductor/electrolyte interfaces 156
W. Jaegermann

IV. PHOTOBIOLOGY

Cytochrome b-559 as a transducer of redox energy into acid–base energy in photosynthesis 169
M. Losada, M. Hervás and J. M. Ortega

Variations in the carotenoid complement of the pigment–protein
complexes of *Rhodospirillum rubrum* 180
    R. M. Lozano and J. M. Ramírez

Towards an analogue of the bacterial photosynthetic reaction centre:
synthesis of an oblique bis-porphyrin system containing a 1,10-phen-
anthroline spacer 189
    S. Noblat, C. Dietrich-Buchecker and J. P. Sauvage

Location and organisation of the chlorophyll-proteins of photosynthetic
reaction centres in higher plants 195
    D. J. Simpson, R. Bassi, O. Vallon and G. Høyer-Hansen

Phosphorylation processes interacting *in vivo* in the thylakoid membranes
from *C. reinhardtii* 210
    F.-A. Wollman and C. Lemaire

Isolation and some properties of photosynthetic membrane vesicles enriched in photosystem I from the cyanobacterium *Phormidium laminosum*
by a non-detergent method 215
    J. L. Serra, J. A. G. Ochoa de Alda and M. J. Llama

Outdoor culture of selected nitrogen-fixing blue-green algae for the
production of high-quality biomass 225
    M. G. Guerrero, A. G. Fontes, J. Moreno, J. Rivas,
    H. Rodríguez-Martínez, M. A. Vargas and M. Losada

Hydrogen peroxide photoproduction by biological and chemical systems   233
    M. A. De la Rosa, M. Roncel and J. A. Navarro

Current topics in the biochemistry of Fe-hydrogenases 241
    W. R. Hagen

List of Participants 251

STABILISATION OF Si PHOTOANODES BY SILICON OXYNITRIDE COATINGS.

ATTEMPTED SYNTHESIS OF BINUCLEAR Co COMPLEXES FOR (PHOTO)ELECTROCHEMICAL REDUCTION OF CO_2

A. Mackor, F. Verbeek, T.P.M. Koster, C.I.M.A. Spee and C.W. de Kreuk

Institute of Applied Chemistry TNO
P.O. Box 108, 3700 AC Zeist, The Netherlands

ABSTRACT

For the stabilisation of Si photoanodes by SiON coatings two routes are followed: the use of sufficiently thin (~ 2.5 nm) coating layers, prepared by etching techniques, so that electron-tunneling may occur and the doping of the thicker (~ 6.0 nm) layers with Pt to make them electronically conductive. The etching route gives limited success in that higher and rather stable photocurrents may be obtained than from photoanodes with thicker coatings. More reproducible results and higher photocurrents (still below 0.1%) occur upon Pt doping in the coating. This work is now in a final stage as specific funds have been used up.
The main emphasis of our EC project is in electrocatalytic CO_2 reduction, where we explore synthetic pathways for novel binuclear Co complexes with tetra-azamacrocyclic ligands and their use in electrochemical measurements and synthesis. Although several routes for ligand synthesis have been briefly described in the literature, we find upon repeated attempts that well-characterised samples are hard to obtain. In one case (bis-dioxo-cyclam 11) we have obtained a pure product. In addtion, the rotating-ring disk electrode has been introduced in this study for obtaining kinetic information on mono-and binuclear complexes. For their use in anhydrous media, water-free compelxes have been prepared.

INTRODUCTION

The photo-electrochemical (PEC) cleavage of water to produce hydrogen and oxygen is still a long-term option for the storage of solar energy. No suitable photoelectrode material has yet been found. We have previously reported about titanates (TiO_2, $SiTiO_3$) and α-ferric oxide as unsuccessful candidates (Mackor et al., 1984a; Tinnemans et al., 1986). Thereupon, we have turned to the mass-produced n-Si, which is unstable as a photoanode. In this paper we report on results, which have been obtained by employing an insulating SiO_xN_y coating of appr. 6.0 nm thickness as a protective layer for n-Si. Two lines of research have been followed:
1) trying to obtain electron-tunneling thin layers via chemical etching techniques
2) doping the original layers with platinum to make them electronically more conductive.

The results are reported in the following section.

Our research on the electrocatalytic reduction of CO_2 is a continuation of previous research, sponsored by the EEC (Mackor et al., 1984b). In the referred study we showed the possibility of lowering the overvoltage for CO_2 reduction by using tetra-azamacrocyclic complexes of Co (and Ni) as electrocatalysts. This is due to the coordination of CO_2 to the metal ion in intermediate steps, whereby the O=C=O bonding is weakened. It was felt that further reduction of this overvoltage is possible if one has two metal centers available at short distances for end-on or side-on coordination. Therefore we have considered the literature on the accessability of ligands and possibly complexes of the type

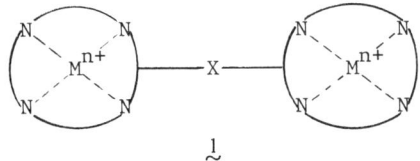

$\underset{\sim}{1}$

in which M^{n+} is a transition-metal ion and X a bridging group, e.g. $(CH_2)_n$ ($1 \leqslant n \leqslant 5$). From this literature three pathways were selected for further study. These results are given in the next-following section.

Finally, we briefly report on the preliminary results of measurements with a rotating ring disk electrode.

STABILISATION OF Si PHOTOANODES BY SiON COATINGS

Results and discussion

While this research is part of the contract, it is mainly reported elsewhere (Mackor et al., 1987a and b). Here the main research lines and results are given.

Two sources of n-Si wafers, coated with 3.7-6.6 nm layers of SiO_xN_y ($x \approx y$), were used with comparable results. From Auger analysis, it follows that upon using these materials as photoanodes in wet PEC cells, the already low photocurrents (of the order of 10^{-8} A/cm^2) decrease upon operation by oxidation of the coated electrode, whereby x increases at the expense of y and by oxidation of Si. The starting SiO_xN_y layers already have a very high resistance ($1.9 \cdot 10^{13}$ $\Omega.cm^{-1}$). This resistivity must be diminished for more efficient operation. Two routes have been followed:

1. Chemical etching by a mixture of HCl, NH_4F and NH_4Cl to obtain a thinner layer with hopefully electron-tunneling properties at layer thickness below 3.0 nm
2. Doping of the coating layer by platinum atoms or clusters for making

them electronically more conductive by so-called resonance-tunneling at
distances of appr. 1.5 nm.

Chemical etching experiments

A homogeneously coated 6.6 nm layer was used and part of the layer was removed by increasing etching times. Two (major) effects have been noted:
a. higher photocurrents, up to a factor 100 higer than the original one in one case, over prolonged periods (hours);
b. poor reproducibility.

Because of the latter effect, approach 1. was abandoned in favour of 2.

Doping SiO_xN_y layers with Pt clusters

A thin layer of Pt was sputtered onto a SiO_xN_y layer (6.6 nm) and thereupon the electrode was annealed at a certain temperature. Even at $300°C$ and short annealing time (30 min) a considerable increase in photocurrent has been routinely obtained at a current density of 14 $\mu A/cm^2$, i.e. appr. 300 times larger than the original photoanode. In some cases an even ten times higher photocurrent density resulted. This is still a factor of 150 too low for practical use (< 0.1%). Our final research in this area is aiming at getting a better understanding of these experiments. These results will be reported at the meeting in more detail.

ATTEMPTED SYNTHESIS OF BINUCLEAR Co COMPLEXES FOR (PHOTO)-ELECTROCHEMICAL REDUCTION OF CO_2

Mononuclear complexes

From previous research (Mackor et al., 1984b), the need had grown to dispose of water-free complexes CoL (or NiL), to be used in anhydrous solvents, e.g. acetonitrile. Two compounds were selected, viz. 2 and 3 (our code numbers MR 794 and VE 2828, resp.).

A previously prepared sample of 2 (Tinnemans et al., 1984) was dissolved in dimethylformamide and it was attempted to remove water by azeotropic distillation. Compound 2 was recovered unchanged, indicating the presence of coordinated water. The synthesis of compound 3 was carried out to give the correct product and elemental analysis. (Tait et al., 1978).

Dinuclear complexes

For the synthesis of dinuclear Co (or Ni) complexes of the type $\underset{\sim}{1}$ the strategy of template synthesis was followed and worked out in two routes. Both principles were described in the literature (Murase et al., 1981, 1983, 1986).

In route I, we synthesized various tetraketones $\underset{\sim}{4}$ (n = 2,3 or 6) and reacted $\underset{\sim}{4a}$ (n = 3) with a tetraamine $\underset{\sim}{5a}$ or $\underset{\sim}{5b}$ in the presence of a cobalt(II) salt in the hope of obtaining binuclear complex $\underset{\sim}{6a}$ or $\underset{\sim}{6b}$.

$\underset{\sim}{6a}$ (15-membered ring)
$\underset{\sim}{6b}$ (14-membered ring) using 2.3.2-tetramine $\underset{\sim}{5b}$

In neither case could we obtain a well-defined and crystalline product. In route II, the building block for a dinuclear ligand was a hexaamine 7, which was reacted with 2,6-diacetylpyridine 8 and $Ni^{II}Cl_2$.

[Scheme: reaction of 2,6-diacetylpyridine 8 with hexaamine 7 and $Ni^{II}Cl_2$]

1. $Ni^{II}Cl_2$; NaOAc/MeOH/H_2O
2. $NaClO_4$

[Scheme: dinuclear Ni complex 9 with 4+ charge and 4 ClO_4^- counterions]

In this case some product was obtained, but the analysis of the product does not fit with pure compound 9. Moreover, unexpected problems occurred during the preparation of hexaamine 7:

$$H_2NCH_2CH_2NH_2 + 4\ CH_2=CHCN \xrightarrow[\text{in acrylonitrile}]{\text{HOAc}}$$

$$(N\equiv CCH_2CH_2)_2NCH_2CH_2N(CH_2CH_2C\equiv N)_2 \quad 80\% \xrightarrow[t = 125°C\ 6\ h]{\substack{200\ \text{bar}\ H_2\\ \text{Raney Ni}}} 7$$

Two successive preparations in an autoclave exploded (or caused the safety plate in the autoclave to break) under these conditions, without personal injuries and without a clear cause. Thereupon the autoclave approach was abandoned and other reducing agents (Redal, $NaAlH_2 \cdot (OCH_2CH_2OCH_3)_2$; B_2H_6/diglyme and B_2H_6/THF) were used without much success. Fortunately, the autoclave method had yielded sufficient product to continue the synthesis. At this point a _third pathway_ was examined to obtain dinuclear complexes (Fabbrizzi et al., 1984, 1986). Coupling of the 2.3.2-tetraamine _5b_ with a tetracarboxylic ester _10_, yielded the bis-dioxocyclam _11_.

_11_

The sample appears to be a mixture if NMR spectra are taken, e.g. one hour after dissolution in D_2O. However, elemental analysis is in good agreement with structure _11_. Moreover, ^{13}C NMR spectra directly taken upon dissolution show only six ^{13}C signals, as expected. Compound _11_ is a potential dinuclear ligand, which is now under study. It has been reduced by B_2H_6 to give the biscyclam _12_, the analysis of which is underway.

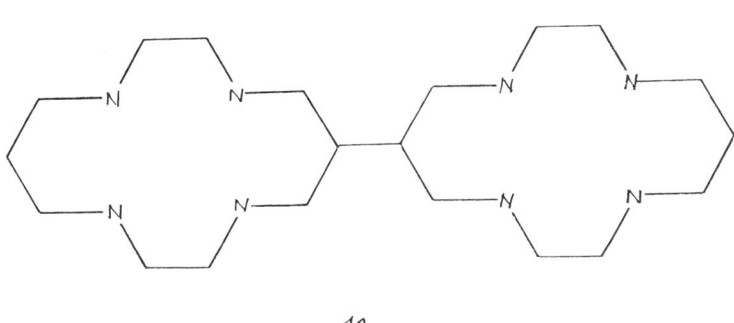

12

In conclusion, the synthesis of dinuclear complexes of Co (or Ni) is cumbersome and it has not yet been achieved. Earlier reports in the literature did not (always) refer to preparative-scale experiments, but rather to spectroscopic characterisation. It seems, however, feasible to obtain complexes of type 1 and therefore these experiments are continued.

RRDE experiments

A rotating ring disk electrode has been manufactured at TNO and it was successfully tested. Preliminary experiments with some old complexes (Tinnemans et al., 1984) showed that these materials had deteriorated over time. The new complexes were either not yet available in the pure state or were intended for use in anhydrous solvent (3 in MeCN), where stability problems occur during electrochemical operation.

ACKNOWLEDGEMENT

The research on water cleavage was supported by PEO, the Netherlands Project Office for Energy Research, by an extended contract, ending on September 30, 1987. The authors are grateful to Dr. W. Visscher of Eindhoven Technical University for supplying to them the drawings of a RRDE.

REFERENCES

Fabbrizzi, L., Forlini, F., Perotti, A. and Seghi, B. 1984. Stepwise incorporation of copper(II) into a double ring octaaza macrocycle and consecutive oxidation to the trivalent state. Inorg.Chem., 23, 807-813.

Fabbrizzi, L., Montagna, L., Poggi, A., Kaden, T.A. and Siegfried, L.C. 1986. Ditopic receptors for transition metal ions: a heterobimetallic nickel(II)-copper(II) bis(macrocyclic) complex and its stepwise oxidation to the tervalent state. Inorg.Chem., 25, 2671-2672.

Koster, T.P.M., Spee, C.I.M.A., De Kreuk, C.W. and Mackor, A. 1987. The stabilization of silicon photoelectrodes by a silicon oxynitride coating. Extended abstracts, 38th Meeting International Society of Electrochemistry, Maastricht, September 13-18, 1987. Vol. II, Abstr. 6.73. Pp.659-661.

Mackor, A., Thewissen, D.H.M.W., Eeuwhorst-Reinten, M., Koster, T.P.M., De Kreuk, C.W., Timmer, K., Tinnemans, A.H.A. and Van der Zouwen-Assink, E.A. 1984a. Homogeneous and heterogeneous photocatalytic cleavage of water using sunlight. Final Report, CEC, Luxembourg, EUR 9527 EN.

Mackor, A., Tinnemans, A.H.A., Koster, T.P.M., De Kreuk, C.W., Thewissen, D.H.M.W. 1984b. Photoelectrochemical reduction of CO_2. Final Report, CEC, Luxembourg, EUR 9529 EN.

Mackor, A., Koster, T.P.M., Spee, C.I.M.A. and De Kreuk, C.W. 1987a. Stabilisation of electrodes in the photo-electrochemical cleavage of water. Condensed annual report 1986 (15 pp; in Dutch). Available upon request.

Mackor, A., Koster, T.P.M., Spee, C.I.M.A. and De Kreuk, C.W. 1987b. Stabilisation of Si photoanodes by SiO_xN_y coating for their use in photoelectrochemical water cleavage. Final report, in preparation. Available in 1988 upon request.

Murase, I., Hamada, K. and Kida, S. 1981. Binuclear copper(II) complexes containing two N_4-macrocyclic rings. Inorg.Chim.Acta $\underline{54}$, L171-L173.

Murase, I., Hamada, K., Ueno, S. and Kida, S. 1983. Binuclear copper(II) and nickel(II) complexes containing two N_4-macrocyclic rings connected with a methylene chain of various lengths. Synth.React.Inorg.Met.-Org.Chem. $\underline{13}$, 191-207.

Murase, I., Ueno S. and Kida, S. 1986. Bis(isocyclam)dicopper(II) complexes with a linear methylene chain bridge. Inorg.Chim.Acta $\underline{111}$, 57-60.

Tait, A.M. and Busch, D.H. 1987. 2,3-Dimethyl-1,4,8,11-tetraazacyclotetradeca-1,3-diene (2,3-Me_2[14]1,3-diene-1,4,8,11-N_4) complexes. Inorg.Synth. $\underline{18}$, 27-29.

Tinnemans, A.H.A., Koster, T.P.M., Thewissen, D.H.M.W. and Mackor A. 1984. Tetraaza-macrocyclic cobalt(II) and nickel(II) complexes as electron-transfer agents in the photo(electro)chemical and electrochemical reduction of carbon dioxide. Recl.Trav.Chim. Pays-Bas $\underline{103}$, 288-295.

Tinnemans, A.H.A., Koster, T.P.M., Thewissen, D.H.M.W. and Mackor, A. 1986. Photoelectrochemical properties of polycrystalline Mg-doped p-type iron(III) oxide. Ber. Bunsenges. Phys.Chem. 90, 383-390.

STUDIES ON ISOLATED PLANT PIGMENT-PROTEIN COMPLEXES

G. Porter and L. Giorgi

Davy Faraday Research Laboratory,
Royal Institution of Great Britain,
21 Albemarle Street, London W1X 4BS, UK

ABSTRACT

The development of processes deriving energy from biomass depends on knowledge of photosynthetic mechanisms, particularly those in the chlorophyll-protein complexes of the light harvesting complex (LHC2), Photosystem 1 (PS1) and Photosystem 2 (PS2). These three pigment-protein complexes (isolated from higher plant thylakoid membranes) have been studied using steady-state and picosecond time-resolved absorption and fluorescence techniques, together with circular dichroism. The principal findings of interest so far are as follows. (1) LHC2: the degree of heterogeneity with respect to the oligomeric state of the LHC2 pigment-protein trimers is extremely sensitive to the detergent:chlorophyll ratio used; the low detergent condition results in extensive aggregation of the trimers with a geometry similar to that found in the 2-D crystals of the LHC2. (2) PS1: energy transfer from the antenna chlorophylls to the P700 trap occurs with a time constant of 15 ± 1 ps. (3) PS2: using a PS2 reaction centre free of antenna chlorophylls, we have been able to observe a transient signal which corresponds to the charge separated radical pair $P680^+$ Pheophytin$^-$.

INTRODUCTION

This Workshop is on Photochemical and Photobiological processes for producing energy rich compounds and more specifically we are part of the sub-programme "Energy from biomass".

Our work in this area is on the primary processes of photosynthesis which occur in the photochemical, i.e. the excited state, regime which, in turn, is confined to the first few nanoseconds of the process.

One may ask two questions about the relevance of this work in the energy field; first, why photosynthesis? And second, why the sub-nanosecond region of times?

The first needs little justification to a group concerned with biomass, which is all derived from photosynthesis, the most successful of all solar energy methods throughout our history. Suffice it to say that whether our eventual target is photosynthesis *in vitro*, or improved photosynthesis *in vivo*, we need to know as much as possible about the mechanism of the natural process.

The answer to the second question, why the study of the very fast primary processes?, is, first of all, because, like Everest, they are there. We need to study all aspects of photosynthesis and it is the one we ourselves have chosen because it involves specialised techniques where we are best equipped to contribute. But one may go further and say that these are the primary steps, the only ones which are true photochemistry and which are, therefore, unique to photosynthesis. The pigment-protein complexes responsible for light harvesting and for the electron transfer sequences within the reaction centres of Photosystem 1 (PS1) and Photosystem 2 (PS2) are the heart of the photosynthetic process.

Originally one visualised a totally synthetic photochemical process, manufactured by chemists, which utilised the knowledge gained from studies of natural photosynthesis, but which was more robust than the green plant as well as being more efficient. And indeed great progress has been made in artificial photosynthesis and quite efficient inorganic models of PS1 and PS2 have been made (we have achieved quantum yields in excess of 60% for both the hydrogen and the oxygen producing steps). But these models are already quite complicated, even in homogeneous solution, and to put them together in a complete photosynthetic system involves not only an amalgam of many chemical substances but a complex morphological structure of membranes, surface catalysts etc. Only triads have so far been synthesised, all involving porphyrins. What we are realising is that nature is able to build far more sophisticated chemical manufacturing units than can the chemist, who may not be able to compete economically.

There seem to be two ways out of this dilemma:

1.   The physico-chemical (*in vitro*) route. As already mentioned this is unlikely to be competitive for the manufacture of biomass (including hydrogen) if it involves making an apparatus with the complexity of the living plant. Furthermore it has to solve the problem of transport of products such as hydrogen, manufactured over very large areas, to a point of collection. Both problems can be avoided if the prime energy provided by solar conversion is electrical.

The most promising *in vitro* route is therefore the production of electrical potential. There are a number of photogalvanic and photoelectrochemical devices which we shall hear about this week, and which present possible solutions to this problem, but they will have to compete with the essentially physical photovoltaic devices, which are looking increasingly promising.

2.   The photobiological (*in vivo*) route.   This aims to modify the existing plants, and to model new ones. It can and will, of course, be used for many plant products other than energy but the application to the efficient and convenient manufacture of biomass is potentially the one of largest scale and importance. The efficiency, for example, of photosynthesis *in vivo*, is less than one tenth of what is theoretically possible.

A better understanding of the photosynthetic units of the *in vivo* system are essential for both the *in vitro* and *in vivo* approaches but if one is looking for practical applications the emphasis of the work changes from one approach to the other. If one is to modify photosynthesis *in vivo* one is almost certainly going to use the methods of genetic engineering so that immediately one needs to know the specific protein complexes involved, their sequences and those of their genomes.

Although research on plants has often been of fundamental importance, from Gregor Mendel to Barbara McClintock, biological and genetic knowledge of plants has fallen far behind that of other living organisms. Yet the possibilities, both for understanding and for commerce, are enormous. Large multinational companies are aware of the potentialities, so much so that most of their work in this field is secret.

There was a second very good reason for doing this work at the present time - it has recently become possible! This is because of the advances in isolation, purification and crystallisation of the key pigment-protein complexes of photosynthetic organisms as well as the improving time resolution and precision of time-resolved spectroscopic techniques in the picosecond and sub-picosecond region.

Outstanding problems

The principal questions to be answered at present are as follows:

1) Why this complex electron transfer structure? What is the purpose of each individual molecule? This is probably fairly well answered now in a general way as a consequence of the work on photosynthetic bacteria, and seems to be mainly a means of preventing back reactions, but much more detailed analysis is needed for PS1 and PS2.

2) The oxygen evolution mechanism of PS2 is totally unknown.

3) Light harvesting does not occur in *in vitro* preparations because of concentration quenching, how is this overcome *in vivo*?

4) At the application level of genetic engineering etc, we need correlations between protein complexes, their function and their genomes.

What is known at present

The present state of the art may be summarised as follows:

1) Everybody knows of the spectacular success of unravelling the structure and kinetics of the photosynthetic bacteria, rhodopseudomonas sphaeroides and viridis; the structure by Deisenhöffer, Hüber and Michel (Deisenhofer et al., 1984) following the isolation and crystallisation by Michel (Michel, 1982) and the picosecond kinetics (which came first) by Rockley, Windsor, Cogdell and Parson (Rockley et al., 1975) and also by Dutton, Rentzepis, Netzel et al. (Netzel et al., 1977).

2) For green plants, chromatography, electrophoresis and similar methods have isolated many pigment-protein complexes and determined their molecular weight and some idea of how they are put together.

3) DNA sequencing of the plant genome is now common and leads immediately to sequences of certain proteins. They can frequently be identified with the pigment protein complexes isolated as described in 2. Remarkable homologies are found e.g. between proteins of the bacterial reaction centres (the M and L sub-units) and the plant PS2 reaction centre (proteins D1 and D2), see Barber (1987) for a review.

4) The only crystalline preparations have been on a water soluble protein from green bacteria (Fenna and Matthews, 1975) and the reaction centres of photosynthetic bacteria (Michel, 1982; Allen and Feher, 1984). Recently, however, ever purer preparations of the plant pigment-protein complexes are being isolated and one of them, the LHC2 complex, has been obtained by our colleague, Werner Kühlbrandt at Imperial College, in two dimensional crystalline form; its tertiary structure has been determined to 7Å by image analysis of electron diffraction (Kühlbrandt, 1984).

5) Our own group (composed of Linda Giorgi, David Klug, Jonathon Ide and Ben Crystall) has made good progress on the three primary plant pigment-protein complexes namely, LHC2 (Ide et al., 1986a, 1986b, 1987), PS1 (Giorgi et al., 1986, 1987) and PS2, some of the results of which will be described below.

TECHNIQUES

The techniques used include steady-state and time-resolved absorption and fluorescence spectroscopy, circular dichroism (carried out by Dr A. Drake at University College, London) and fluorescence polarisation. The picosecond fluorescence apparatus (see Ide et al., 1987) is standard for single photon counting and has a time resolution of approximately 120ps (limited by the instrument response function). The picosecond absorption apparatus (see Gore et al., 1986) also utilises techniques which are becoming almost standard today: modelocked picosecond and sub-picosecond pulse trains, amplification, split beam to form pulse and probe and white light generation for polychromatic probe pulses. Most of the absorption work to be described here had a time resolution of about 10ps; we are now using pulse compression to reduce our pulse width to 500fs.

RESULTS

Interesting results have been obtained on each of the three principal plant pigment-protein complexes.

1) The light harvesting complexes were our earliest interest, using the whole chloroplast and not very well defined fragments of it, as well as the light harvesting complex from porphyridium cruentum which provided our most satisfactory results until recently. Now the crystalline preparations of LHC2 by Kühlbrandt have given us much better characterised material to work on.

2) PS1, containing still part of its own light harvesting complex, has provided excellent kinetic results on the energy transfer to the reaction centre, with the subsequent photooxidation of P700, and the time of this whole process.

3) Most exciting of all are the PS2 reaction centres. Because of the overlap of the light harvesting chlorophyll and the reaction centre chlorophyll at 680nm, in the PS2 complex, we

were not at first able to resolve spectrally their kinetic absorption changes. Isolation of this particular complex quite free of light harvesting chlorophyll has transformed the situation and we have been able to observe time-resolved spectra of the primary electron transfer.

The light harvesting complex LHC2

The LHC2, which is mainly associated with PS2, has three main functional roles: (1) to harvest and transfer solar energy to the reaction centres of PS2; (2) to control the distribution of absorbed quanta between the two photosystems under fluctuating external light quality and (3) to mediate interactions between the thylakoid membranes, e.g. the formation of thylakoid stacks. In this work we investigate the effect of changing the detergent:chlorophyll ratio on the resolubilised LHC2 energy transfer kinetics. Our results have been interpreted in light of the known geometry of LHC2 in 2-dimensional crystals (Kühlbrandt, 1984) : a 3-fold axis runs vertically through the centre of each complex, indicating that the LHC2 is a trimer composed of 3 identical sub-units in symmetry-related positions; each monomeric sub-unit consists of a polypeptide which binds a total of 6-11 chlorophyll molecules (specifically, chlorophyll-a and chlorophyll-b molecules).

We have found that the degree of heterogeneity with respect to the oligomeric state of the pigment-protein trimers is dependent upon the detergent:chlorophyll ratio used. Low detergent:chlorophyll ratios result in extensive aggregation of the trimers with a geometry similar to that found in 2-D crystals of the LHC2. Moderate detergent conditions yield predominantly non-aggregated trimers. Excess detergent conditions results in considerable chromophore heterogeneity with protein denaturation through an initial break up of the trimer geometry. These conclusions have been reached from a variety of experimental evidence. The pertinent data are given below.

The fluorescence lifetime data reveals (^{1}/e) lifetimes of 3.53 ($\pm$0.04)ns and 1.10 ($\pm$0.01)ns for a stable, efficiently energy-transfering state of the LHC2 (namely, under moderate detergent conditions). Subnanosecond lifetimes are observed under conditions leading to aggregation while a long component of 5.50 ($\pm$0.16)ns corresponding to free Chl-a is found when the detergent:chlorophyll ratio is high. The circular dichroism shows a major Chl-b exciton, a Chl-a/b exciton and a further 'quenching' Chl-b exciton. These have been attributed respectively to: a C_3 symmetric Chl-b interaction for which the intact C_3 protein trimer geometry is a prerequisite; a dimeric Chl-a/b interaction, the presence of which is critically dependent on the detergent type and a further Chl-b interaction which arises from the presence of aggregated trimers.

From our results we believe that, *in vitro*, the minimum stable functional unit corresponds to a C_3 symmetric pigment-protein trimer complex.

Reaction centres of Photosystem 1

Briefly, the sequence of electron carriers within the PS1 reaction centre can be summarised as:

$$P700 \ldots A_0 \ldots A_1 \ldots A_2 \ldots P430 \text{ (iron-sulphur centres A and B)}$$

P700 is the primary electron donor, A_0 may be a chlorophyll-a, A_1 a quinone and A_2 (also called X) a specialised iron-sulphur centre. A and B are bound iron-sulphur centres characterised by absorption at 430nm.

Experiments have been carried out on PS1 reaction centres, isolated from pea chloroplasts, with a chl/P700 ratio of 50 (see Giorgi et al., 1987). P700 was chemically reduced in the dark by ascorbate and chemically oxidised by ferricyanide. The time-resolved absorption spectra obtained are shown in Fig.1. Two obvious spectral features occur, one centred at 690nm and one at 700nm.

The feature at 690nm is dominant at early times and appears in the spectra of samples with P700 either chemically oxidised or chemically reduced. This suggests that the 690nm signal can be attributed to the excitation of antenna chlorophyll molecules to the singlet state. It decays with a lifetime of 15 ± 1ps (obtained from global analysis of the spectra from t = 6.5ps to t = 270ps, without deconvolution: this yields a good fit to the data with a single exponential decay). The decay of the 690nm signal is, within the resolution of these experiments (10ps) independent of the redox state of P700. The 690nm feature undergoes a blue shift as it decays, finally being centred at 675nm. Previous experiments (Giorgi et al., 1986) have shown that the residual bleach at 675nm decays with a lifetime of between 1 and 2ns; it is believed that this signal is due to residual triplet chlorophyll molecules which are quenched by carotenoid molecules within the antenna. The 690nm signal cannot be explained solely as a combination of ground state bleach and stimulated emission from the singlet chlorophyll. Instead, it is indicative of sub-picosecond energy transfer, within the antenna, to core chlorophyll molecules that absorb and emit significantly redder than the majority of chlorophyll molecules.

The 700nm spectral feature is much narrower than the 690nm signal and is only observed in the spectra of samples with P700 chemically reduced. This suggests that it can be attributed to the photooxidation of P700 molecules. By studying the difference between the oxidised and reduced spectra, at each time delay, it is possible to follow the rise of the P700 signal as the 690nm signal decays. Despite the short antenna decay lifetime and the spectral overlap of the 690nm and P700 signals, it is clear that the P700 signal only appears once the excitation pulse is over i.e. it is delayed relative to the maximum of the 690nm signal. The rise time of the P700 signal is similar to the decay time of the 690nm signal, with the P700 signal reaching 80% of its final value 20 ± 3ps after the maximum of the excitation pulse.

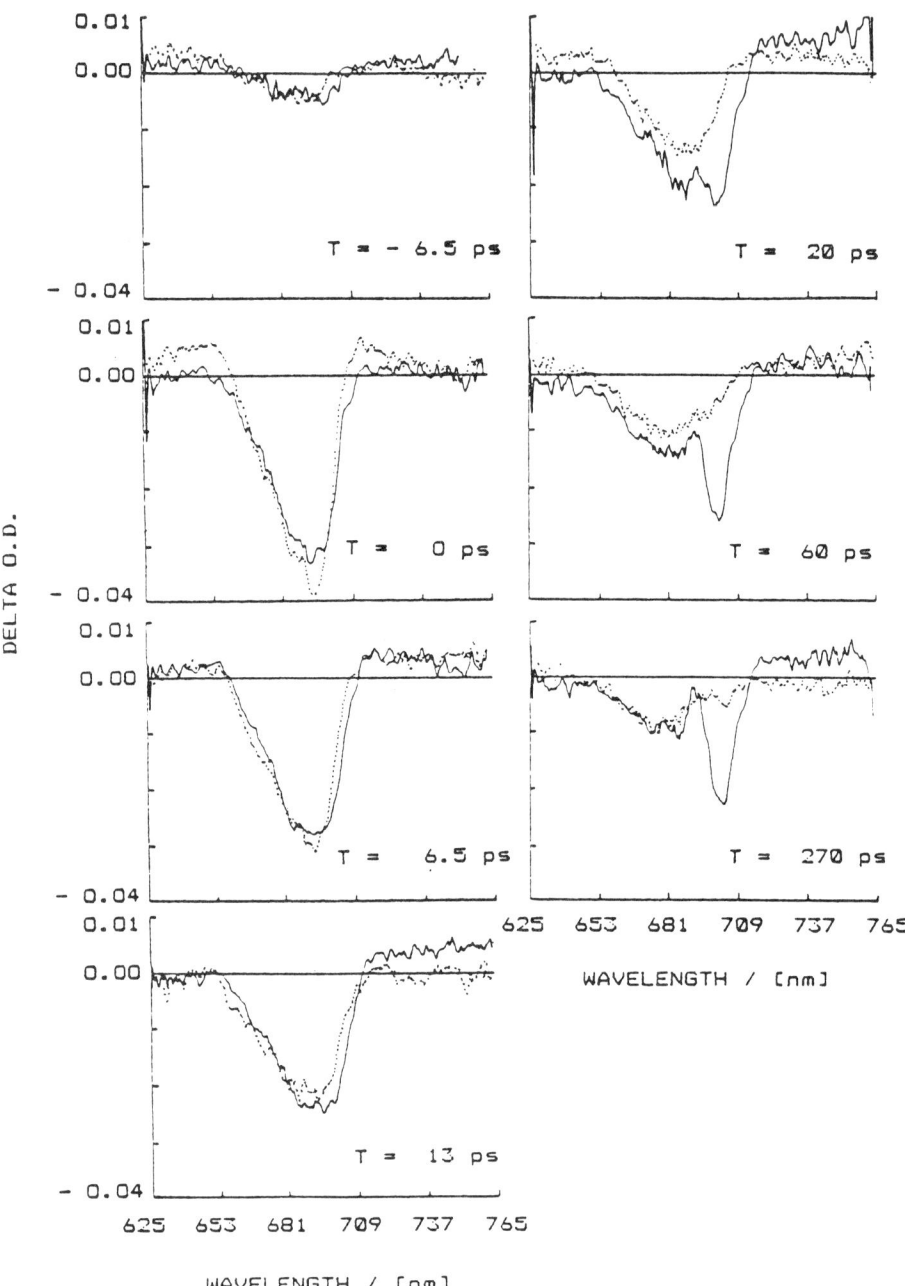

Fig. 1 Transient absorption spectra of PS1 reaction centres. (———) P700 chemically reduced; (......) P700 chemically oxidised.

Note, throughout these experiments it has been necessary to eliminate multiple excitation of single antennae which result in annihilation and rapid initial decay of the bleach. This is not observed if the magnitude of the initial 690nm signal is reduced to less than 0.1 OD.

Also, although we could clearly identify the photooxidation of P700, we have found no evidence for the existence of a chlorophyll molecule acting as a primary acceptor in PS1. This is in line with the work on photosynthetic bacteria where it has been shown that the related molecule (a monomeric bacteriochlorophyll) either does not receive an electron, or that it loses the electron at a faster rate than that at which it is gained.

Reaction centres of Photosystem 2

The sequence of electron carriers within the PS2 reaction centres can be summarised as:

$$P680 \ ... \ Ph \ ... \ Q\text{-}Fe$$

P680 is the primary electron donor, Ph is pheophytin-a and Q-Fe is a plastoquinone-iron complex.

Our initial experiments with PS2 reaction centres were plagued with the problem that the antenna bleach and the reaction centre spectral change both occur at 680nm. However it is now possible to isolate PS2 reaction centres free of antenna chlorophylls (Nanba and Satoh, 1987). This new PS2 complex is thought to be the PS2 reaction centre. It contains 4 chlorophyll-a molecules, 2 pheophytin-a molecules, 1 cytochrome b-559 and some ß-carotene. It contains no plastoquinone, indicating that both the secondary acceptor Q_A and the secondary donor Z must be lost. In the absence of secondary donors and acceptors, the photochemical activity of the complex should be limited to the formation of the primary radical pair $P680^+$ Pheophytin$^-$. Fig.2 shows the absorption difference spectra of these PS2 reaction centres at 0ps (pump and probe pulses overlapped) and at +6ps. The 0ps absorption spectrum can be totally attributed to loss of ground state absorption by the complex, following its excitation. This 0ps spectrum can be directly compared with the ground state absorption spectrum of the PS2 reaction centre obtained using our picosecond spectrometer. The peak at 415nm is attributable to pheophytin-a absorption and that at 430nm is attributable to chlorophyll-a absorption. To interpret the spectrum at 6ps it is necessary to simulate the spectrum one would expect to observe at 6ps for the primary radical pair $P680^+$ Pheophytin$^-$: it will be the sum of the difference spectrum for Pheophytin$^-$ and that for $P680^+$. It becomes clear when doing this that we are indeed observing the primary radical pair $P680^+$ Pheophytin$^-$, in these PS2 reaction centres, and that it is being formed within 6ps of flash excitation. Only one other measurement of this sort, on this particular new PS2 complex has been reported in the literature. In this paper (Danielius et al., 1987), the authors conclude also that the spectral features they observe are attributable to the formation of $P680^+$

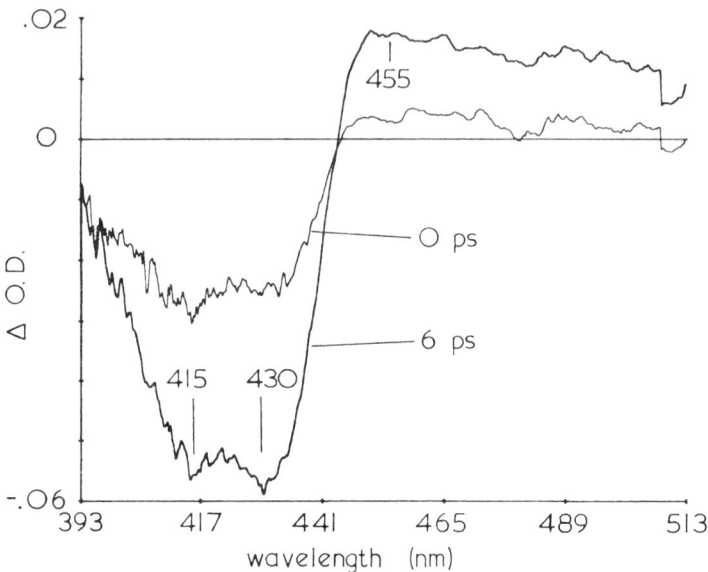

Fig. 2 Transient absorption spectra of PS2 reaction centres.

Pheophytin⁻. However, their time resolution is of the order of 25ps and the spectrum they present is at 4ns after excitation. Hence the best estimate that they are able to give for the rate of formation of the primary radical pair is that it occurs in less than 25ps.

REFERENCES

Allen, J.P. and Feher, G. (1984) Proc. Natl. Acad. Sci. USA 81, 4795-4799
Barber, J. (1987) TIBS 12, 321-326
Danielius, R.V., Satoh, K., van Kan, P.J.M., Plijter, J.J., Nuijs, A.M. and van Gorkom, H.J. (1987) FEBS Lett. 213, 241-244
Deisenhofer, J., Epp, O., Miki, K., Huber, R. and Michel, H. (1984) J. Mol. Biol. 180, 385-398
Fenna, R.E. and Matthews, B.W. (1975) Nature 258, 573-577
Giorgi, L.B., Doust, T., Gore, B.L., Klug, D.R., Porter, G. and Barber, J. (1986) Biochem. Soc. Trans. 14, 47-48
Giorgi, L.B., Gore, B.L., Klug, D.R., Ide, J.P., Barber, J. and Porter, G. (1987) in Progress in Photosynthesis Research (Biggins, J., ed.) Vol. 1, pp 257-260, Martinus Nijhoff Publishers
Gore, B.L., Doust, T.A.M., Giorgi, L.B., Klug, D.R., Ide, J.P., Crystall, B. and Porter, G. (1986) J.C.S. Faraday Trans. 2 82, 2111-2116
Ide, J.P., Klug, D.R., Kühlbrandt, W., Giorgi, L.B., Porter, G., Gore, B.L., Doust, T. and Barber, J. (1986a) Biochem. Soc. Trans. 14, 34
Ide, J.P., Klug, D.R., Crystall, B., Gore, B.L., Giorgi, L.B., Porter, G., Kühlbrandt, W. and Barber, J. (1986b) J.C.S. Faraday Trans. 2 82, 2263-2266
Ide, J.P., Klug, D.R., Kühlbrandt, W., Giorgi, L.B. and Porter, G. (1987) Biochim. Biophys. Acta, in press
Kühlbrandt, W. (1984) Nature 307, 478-480

Michel, H. (1982) J. Mol. Biol. 158, 567-572
Nanba, O. and Satoh, K. (1987) Proc. Natl. Acad. Sci. USA 84, 109-112
Netzel, T.L., Rentzepis, P.M., Tiede, D.M., Prince, R.C. and Dutton, P.L. (1977) Biochim. Biophys. Acta 460, 467-478
Rockley, M.G., Windsor, M.W., Cogdell, R.J. and Parson, W.W. (1975) Proc. Natl. Acad. Sci. USA 72, 2251-2255

PRIMARY REACTIONS IN PLANT PHOTOSYNTHETIC REACTION CENTERS

P. Mathis

Service de Biophysique, Département de Biologie, CEN/Saclay
91191 Gif-sur-Yvette cedex, France

ABSTRACT
Primary steps of photoinduced electron transfer have been studied in plant reaction centers (PS-I and PS-II), by flash absorption and EPR. In PS-I two questions were investigated : i) the properties of the primary radical pair $P-700^+$, A_0^- (kinetics of decay ; nature of A_0, presumably a specialized chlorophyll $\bar{a}$; decay by recombination to populate the P-700 triplet state) and ii) the nature of the secondary acceptor A_1. Extraction-reconstitution experiments indicate that A_1 is very probably a molecule of vitamin K_1.
The core of the PS-II reaction center has been prepared. Flash absorption showed that this core is able of efficient charge separation to form the primary radical pair which decays in about 30ns. The recombination populates the P-680 triplet state, which does not transfer to beta-carotene and can be detected by spin-polarized ESR. The yield of formation and kinetics of decay of the radical-pair have been measured in various PS-II preparations. The data are in favor of an equilibrium between the radical-pair and chlorophyll excited state in the antenna.

INTRODUCTION

Photosynthetic reaction centers efficiently transform the energy of light into a form of chemical energy (Mathis and Rutherford, 1987). These reaction centers are multiproteic assemblies located in the photosynthetic membranes. Following absorption of light by appropriate pigments, excitation energy is transferred to a species labelled P (also named primary electron donor) which gets electronically excited and transfers an electron to a primary electron acceptor, triggering a sequence of electron transfer reactions. In less than 1ms after the triggering event, about 50% of the photon energy is stored in the form of redox energy. In plant photosynthetic membranes, two types of reaction centers (named PS-I and PS-II) cooperate in electron transfer. In this work, we use a conjonction of biochemical and of spectroscopic methods to progress in understanding the structure and composition of these reaction centers, and the factors which permit a rapid and efficient charge separation.

MATERIALS AND METHODS

Particles enriched in the PS-I reaction center were prepared from spinach thylakoid by treatment with SDS (Brettel and Sétif, 1987) or with digitonin and then with water-saturated diethylether (Sétif et al, 1987). Particles were also prepared from the cyanobacterium Synechocystis 6803 by treatment with octylglucoside and then eventually with a hexane-methanol mixture (Biggins and Mathis, 1987). The PS-II reaction center core was prepared by action of Triton X-100 on spinach membranes, according to Nanba and Satoh (1987). Other PS-II preparations were made according to published procedures, as reported by Hansson et al (1987).

Flash absorption kinetics were measured following excitation of the biological material with either a picosecond laser pulse (532nm, 20ps) or with a nanosecond pulse (595nm, 10ns). The absorption changes were measured with silicon photodiodes, amplified and recorded as described by Brettel and Sétif (1987) or by Takahashi et al (1987). Low temperature work was done with a cryostat cooled with liquid helium.

Electron spin resonance spectra were recorded in darkness or under illumination, at low temperature, with a Bruker ER 200D-SRC spectrometer, as described by Frank et al (1987) or by Sétif et al (1987).

PHOTOSYSTEM-I REACTION CENTER

Our research aims at elucidating the chemical nature of the redox centers involved in light-induced charge separation, the mechanisms of reaction and the role of the protein counterpart. During this last year our work has concerned mainly the primary radical pair and the secondary acceptor A_1.

a. Primary radical pair. The primary partners consist of the pair (P-700, A_0) which reacts as follows : (P-700, A_0) $\rightarrow$ (P-700*, A_0) $\rightarrow$ (P-700$^+$, A_0^-). Under normal conditions, this primary radical pair evolves very quickly (subnanosecond) by electron transfer from A_0^- to the next acceptor A_1, and then by reduction of P-700$^+$ by the secondary donor, the copper protein plastocyanin. The secondary reactions can be inhibited by several biochemical manipulations (mild treatment with SDS, extraction of A_1 by ether-water or by hexane-methanol). Under these conditions, flash absorption studies allowed us to studies several properties of the radical pair.

One of our objectives was to identify the primary acceptor A_0. The state ($P\text{-}700^+$, A_0^-) was found to decay with $t\frac{1}{2}$ = 45ns (at 278K ; about 80ns at 10K). By flash absorption with nanosecond time resolution, we measured the difference spectrum due to the formation of the radical pair. Subtracting the $P\text{-}700/P\text{-}700^+$ difference spectrum, we have obtained the A_0/A_0^- difference spectrum (Mathis et al, 1987). This one includes bleachings around 430 and 690nm, and broad positive bands in the blue (460-500nm) and the near infra-red. This spectrum well fits the hypothesis that A_0 is a specialized molecule of chlorophyll a or a dimer of it. The bleaching at 690nm differs from a bleaching at 670nm reported by several groups in photoaccumulation experiments. We propose that A_0^- is not stable and under steady state illumination reduces neighbour chlorophyll a molecules absorbing at 670nm.

The kinetics of decay of ($P\text{-}700^+$, A_0^-) are due to two mechanisms of charge recombination, going back to the ground neutral state ($P\text{-}700$, A_0) or to the triplet state ($^3P\text{-}700$, A_0). The triplet state is formed with a temperature-dependent yield : 30% at 278K, 75% at 10K. Its decay also is temperature-dependent, with a half-time of 6-10us at 278K and 800us at 10K. We have found, however, that treatment of PS-I with organic solvents significantly changes the decay kinetics : at 278K the decay is slower (half-time 30us) and highly oxygen-dependent : 30us under air, 600us without oxygen (Sétif et al, 1987). We propose that $^3P\text{-}700$ is normally quenched by triplet-triplet energy transfer to carotenoids, which is faster than the reaction with oxygen. Carotenoids are removed by organic solvents and $^3P\text{-}700$ then decays much more slowly, permitting to observe a quenching by oxygen. The amount of $^3P\text{-}700$ which is formed in the radical-pair recombination can be varied by application of a weak magnetic field, following a prediction of the radical-pair mechanism (Brettel and Sétif, 1987). The most interesting property found in PS-I is that the triplet yield first increases when the magnetic field increases, achieving a maximum for H=60G, and then decreases slowly at higher field. This behaviour permit to determine an exchange interaction of 60G between $P\text{-}700^+$ and A_0^- (it is argued that the major interaction is by the exchange mechanism rather than dipolar). This interaction appears to be much larger than that reported in bacterial reaction centers (about 20G), a difference which probably reflects significant differences in the relative positioning of the primary electron carriers.

b. Chemical nature of the secondary acceptor A_1

In a previous study we have found that, at low temperature, PS-I electron transfer is largely blocked away from A_1^-, and that the state ($P-700^+$, A_1^-) decays with a half-time of 130us. Analysis of the absorption spectrum of that state showed that A_1^- is presumably a quinone radical anion (Brettel et al, 1986). Chemical analysis, following separation by HPLC, has shown that phylloquinone (a naphthoquinone also named vitamin K_1) is the only quinone present in PS-I. We have found 2 moles of phylloquinone per PS-I. Extraction with dry hexane does not change the electron transfer reactions ; this treatment only extracts only one phylloquinone per PS-I (Biggins and Mathis, 1987).

We have studied PS-I particles treated so as to completely remove phylloquinone (Ikegami et al, 1987 ; Sétif et al, 1987 ; Biggins and Mathis, 1987). In flash absorption studies, at room temperature, these preparation behave as if A_1 were totally absent : the primary-radical pair decays in the nanosecond time range, largely via the triplet route. Forward electron transfer to the iron-sulfur centers seems to be nearly absent. Readdition of pure phylloquinone reconstitutes forward electron transfer as shown by the kinetics of flash-induced absorption changes and by the ability to photoreduce NADP (with addition of ferredoxin and ferredoxin-NADP-reductase). The reconstitution is rather specific of phylloquinone, and it takes place at a nearly stoechiometric quinone concentration (about 2uM with about 1uM of reaction centers). Another naphthoquinone, vitamin K_3, which has a proton instead of the long side chain found in phylloquinone, is apparently also able to accept electrons from A_0, as shown by nanosecond flash kinetics. The efffect of vitamin K_3, however, is totally reversed by addition of dithionite, whereas the effect of phylloquinone is not affected by dithionite. This shows that phylloquinone binds at the A_1 site, where it acquires a very low redox potential (the redox potential of the A_1 - A_1^- couple is as low as - 0.9V versus NHE), whereas added vitamin K_3 remains entirely reducible by dithionite and thus does not reconstitute correctly the A_1 function. Experiments are in progress to learn which properties of quinones are required for binding correctly at the A_1 site.

In these studies it appeared that low-temperature photoreduction of the bound iron-sulfur centers is not greatly decreased (about 2x) by the removal of phylloquinone (Sétif et al, 1987). Low temperature photochemistry was measured by three methods : the total amount of iron-sulfur centers A and B reduced by continuous illumination at 77K, the extent of reduction of these centers by saturating laser flashes at 77K, and the flash-induced formation of triplet P-700. These methods show that all the reaction centers are still able to oxidize P-700 and to reduce the iron-sulfur centers. This observation raises serious questions concerning the role of phylloquinone or the significance of low-temperature photochemistry. For the moment we consider that room temperature data are more reliable in indicating that A_1 is a phylloquinone.

PHOTOSYSTEM-II REACTION CENTER

In the recent years, our work on the PS-II reaction center has mainly involved EPR spectroscopy, to identify and understand the functional properties of the secondary acceptors, essentially two quinones Q_A and Q_B interacting with a Fe^{2+} iron, and of the manganese enzyme which realizes the oxidative cleavage of water (see e.g. Beijer and Rutherford, 1987 or Styring and Rutherford, 1987). The recent development of a method permitting to simply isolate a PS-II reaction center core (Nanba and Satoh, 1987) allowed us to study the primary radical-pair (P-680$^+$, I$^-$). We first worked in collaboration with the Japanese group, then developed our own preparation (Hansson et al., 1987 ; Frank et al., 1987 ; Takahashi et al., 1987) and also collaborated with the group of Professor Barber (London, UK).

Flash excitation of the PS-II reaction center core leads to nanosecond absorption changes due to the formation of the primary radical-pair, which we could identify as made of the oxidized primary donor (P-680$^+$) and of the reduced primary acceptor (I$^-$, pheophytin anion radical). The decay half-time is 30ns (slightly longer at low temperature). It is formed with a rather high quantum efficiency (60%). One route of decay leads to the formation of the P-680 triplet state (as shown by flash absorption spectra and kinetics), according to the radical-pair mechanism (as evidenced by the the ESR spectrum of ^{3}P-680, which has the expected a,e,e,a,a,e polarization). The reaction center contains one beta-carotene molecule, and it is remarquable that ^{3}P-680 does not appear to be significantly quenched by carotene. A low-yield formation of ^{3}Car could be evidenced, but probably by a mechanism which does not involve ^{3}P-680.

The biradical formation is not influenced by the redox poising of the medium, in agreement with the reported lack of quinones bound to the complex. However, the complex includes the polypeptides D_1 and D_2 which have presumably the binding sites for Q_A and Q_B. We thus have attempted to reconstitute the Q_A function with various quinones (plastoquinones or analogs with a different isoprenoid chain) and using several protocols (addition of lipids, detergent exchange, etc.). In no case have we been able to reconstitute the Q_A function at the present time. It is possible that the binding site is denaturated, or that binding requires other polypeptides such as CP47 or CP43. The results demonstrate that the photochemical core of PS-II is made of the D_1 and D_2 polypeptides, in agreement with hypotheses based on the analogies between the PS-II and bacterial reaction centers, and on the homology between D_1, D_2 and the L, M polypeptides of bacteria. Cytochrome b_{559} is strongly associated with the complex, but it does not seem to participate in electron transfer.

Illumination of the complex under reducing conditions, at 278K, leads to the photoreduction of pheophytin (Nanba and Satoh, 1987). Illumination at low temperature (200K) allowed us to reduce pheophytin and to trap the material for ESR studies at 20K. The ESR spectrum indicates two species which are tentatively interpreted as reduced pheophytin and an oxidized electron donor. The putative donor spectrum has a high g value (2.0045) and it could be a modified state of the physiological donor Z (Frank et al., 1987). If this interpretation is correct, it would mean that Z is carried by the D_1, D_2 couple, like the other partners in primary photochemistry.

PHOTOSYSTEM-II REACTION CENTER : COUPLING WITH THE ANTENNA

The rather long lifetime of the radical-pair in the reaction center core (30ns) led us to examine its lifetime in other PS-II preparations. It was hypothesized that an equilibrium was established between the radical-pair and the singlet excited state of P-680 or of chlorophyll a in the antenna (we assume that the antenna is made of n chlorophyll molecules) :

$(Chl_n)^* P\text{-}680, I \rightleftharpoons Chl_n P\text{-}680^*, I \rightleftharpoons Chl_n P\text{-}680^+, I^-$

The establishment of such an equilibrium should have two effects after a short flash : the apparent lifetime of the radical-pair should vary with n (because of the 5ns lifetime, at most, of Chl^* and of the occurrence of other traps in the antenna), and the amount of radical-pair should decrease while excitation resides in the antenna. To check that hypothesis we have varied $\underline{n}$, measuring the amount of biradical state and its lifetime in several PS-II preparations with a different antenna size ($n=5$ to 200) (Hansson et al., 1987).

With each type of PS-II membrane we have measured the absorption increase at 820nm induced by a 20-ps laser flash, first under oxidizing conditions : the signal then is due to $P-680^+$ in all the reaction centers. The measurement was then effected under the same conditions, except that Q_A was reduced : the signal then is due to $(P-680^+, I^-)$ or to excited chlorophyll. The results are as follows ($\underline{n}$ is the number of chlorophylls per P-680 or antenna size ; $t_\frac{1}{2}$ is the half-time of the signal obtained when Q_A^- is reduced ; $\underline{b}$ is the ratio $(P-680^+, I^-/P-680^+$ under reducing conditions, i.e. the fraction of reaction centers which are in the radical pair state ; $1-\underline{b}$ is thus the fraction of reaction centers where chlorophyll is excited ; alternatively, $\underline{a}$ is the same ratio if we suppose that there is no excited chlorophyll detected in our experiments) :

- Reaction center core : $n=5$; $t_\frac{1}{2}=30 - 32$ns
- Spinach core complex : $n=60$; $t_\frac{1}{2}= 25$ns ; $\underline{a}=0.62$; $\underline{b}=0.33$
- Chlamydomonas core complex : $n=50$; $t_\frac{1}{2}=14$ns ; $\underline{a}=0.67$; $\underline{b}=0.41$
- Oxygen-evolving spinach core complex : $n=66$; $t_\frac{1}{2}=2$ns (50%), 10ns (50%) ; $\underline{a}=0.64$; $\underline{b}=0.36$
- Chlamydomonas membranes : $n=115$; $t_\frac{1}{2}=4$ns ; $\underline{a}=0.60$; $\underline{b}=0.29$
- Spinach membranes : $n=200$; $t_\frac{1}{2}=3$ns ; $\underline{a}=0.55$; $\underline{b}=0.20$.

The data first show that the half-time largely varies within our set of biological materials. In general it is shorter in particles with a larger antenna, in agreement with the initial hypothesis. However the half-time remains long (15-25ns) in PS-II preparations with a rather large antenna (about 60 chlorophylls) ; it seems to decrease sharply in the "membranes", which are the most intact preparations. So we cannot really decide whether the important parameter is the antenna size or the intactness of the structure. The quantitative parameters $\underline{a}$ and $\underline{b}$ also indicate a trend toward less radical-pair in larger particles ; the trend is naturally more pronounced in the $\underline{b}$ hypothesis. Altogether it appears that the data are in favor of the hypothesis under study of a rather fast equilibrium between excited state of chlorophyll in the antenna and radical pair in the reaction center, although other parameters will also have to be considered.

REFERENCES

Beijer, C. and Rutherford, A.W. 1987. The iron-quinone acceptor complex in Rhodospirillum rubrum chromatophores studied by EPR. Biochim. Biophys. Acta 890, 169-178.

Biggins, J. and Mathis, P. 1987. The functional role of vitamin K_1 in the Photosystem I of the cyanobacterium Synechocystis 6803 (submitted).

Brettel, K., Sétif, P. and Mathis, P. 1986. Flash-induced absorption changes in photosystem I at low temperature : evidence that the electron acceptor A_1 is vitamin K_1. FEBS Lett. 203, 220-224.

Brettel, K. and Sétif, P. 1987. Magnetic field effects on primary reactions in Photosystem I. Biochim. Biophys. Acta (in press).

Frank, H.A., Hansson, O. and Mathis, P. 1987. Low temperature ESR and absorption spectroscopy of the Photosystem II reaction center complex (submitted).

Hansson, O., Duranton, J. and Mathis, P. 1987. Yield and lifetime of the primary radical pair in preparations of Photosystem II with different antenna size. Biochim. Biophys. Acta (in press).

Ikegami, I., Sétif, P. and Mathis, P. 1987. Absorption studies of Photosystem I photochemistry in the absence of vitamin K_1. Biochim. Biophys. Acta (in press).

Mathis, P. and Rutherford, A.W. 1987. The primary reactions of Photosystems I and II of algae and high plants. In "Photosynthesis" (Ed. J. Amesz). (Elsevier, Amsterdam) pp. 63-96.

Mathis, P., Ikegami, I. and Sétif, P. 1987. Nanosecond flash studies of the absorption spectrum of the Photosystem I primary acceptor A_0 (submitted).

Nanba, O. and Satoh, K. 1987. Isolation of a Photosystem II reaction center consisting of D_1 and D_2 polypeptides and cytochrome b-559. Proc. Natl. Acad. Sci. USA, 84, 109-112.

Sétif, P., Ikegami, I. and Biggins, J. 1987. Light-induced charge separation in Photosystem I at low temperature is not influenced by vitamin K_1. Biochim. Biophys. Acta (in press).

Styring, S. and Rutherford, A.W. 1987. In the Oxygen-evolving complex of Photosystem II the S_0 state is oxidized in the S_1 state by D^+. Biochemistry, 26, 2401-2405.

Takahashi, Y., Hansson, O., Mathis, P. and Satoh, K. 1987. Primary radical pair in the Photosystem II reaction centre. Biochim. Biophys. Acta, 893, 49-59.

IMMOBILIZED PHOTOSYNTHETIC SYSTEMS FOR THE PRODUCTION OF
AMMONIA AND PHOTOCURRENTS

D.O. Hall, K.K. Rao, H. de Jong, M. Gratzel*
M.C.W. Evans**

Department of Biology, King's College London (KQC),
Campden Hill Road, London W8 7AH, U.K.
*Institute de Chimie Physique, E.P.F.L.,
1015 Lausanne, Switzerland.
**Department of Botany and Microbiology,
University College London, Gower Street, London WC1E 6BT, U.K.

ABSTRACT

Cyanobacteria immobilized in polymer foam supports were used for the study of ammonia production in photobioreactors. Immobilization resulted in the stabilisation and enhancement of nitrogenase and hydrogenase activities compared to these enzyme activities in the free-living cells. In Anabaena azollae the heterocyst frequency of the immobilized cells doubled relative to the free-living cells and reached a level of 14-17%. Ammonia photoproduction by the immobilized cells in the presence of the glutamine synthetase inhibitor L-methionine-D,L-sulphoximine (MSX) was 5 to 10 fold higher than that from free-living cells. Continuous ammonia production lasting for more than a month, at rates of 300 umole/l.h, was observed in packed-bed flow reactors under 12h light/dark cycles and with intermittent MSX pulses. We have now designed a photobioreactor in which immobilized A.azollae can produce ammonia at rates of ca. 1,000 umoles/l.h. Low temperature scanning electron microscopy revealed that the surface of the cells were covered in a layer of hydrated mucilage in A.azollae immobilized in polymer foam pieces and in its natural symbiotic niche in the Azolla fern.

Photosystem (PS) I and PSII particles deposited on naked, doped, or derivatised TiO_2 electrodes were used as photoanodes in regenerative, dye-sensitized photoelectrochemical cells. Spinach PSI deposited on a TiO_2 electrode, in the presence of ascorbate as PSI electron donor, on illumination with white light generated photocurrents of the order of 150 to 250 uA. The photon to current conversion efficiency approached 0.3%. Spinach and pea PSII particles deposited on TiO_2 electrodes generated continuous photocurrents of 5 to 15 uA with water as electron donor and dimethyl benzoquinone as electron acceptor: the efficiency was less than 0.1%. The reaction was inhibited by DCMU. Preparation of cyanobacterial PSI particles when examined by ESR spectroscopy showed signals identical to those assigned to Fe-S centres A1 and A0 in higher plant PSI.

INTRODUCTION

Cyanobacteria are O2-evolving photosynthetic prokaryotes, many of which are able to fix atmospheric N2 via an ATP-dependent nitrogenase activity. The nitrogenase enzyme is oxygen-sensitive; however, it is localized in specialized cells called heterocysts which lack PSII and has an envelope impermeable to O2. Under normal physiological conditions, the nitrogen fixed in the cells as ammonia is rapidly transformed to glutamine, a reaction catalysed by the enzyme glutamine synthetase. The glutamine subsequently enters the cellular metabolic pool [Fig.1]. The cells can be induced to excrete ammonia by addition of L-methionine-D,L-sulfoximine (MSX), a glutamine analogue which inhibits glutamine synthetase. Since glutamine is an essential metabolite, to maintain cell growth, the supply of MSX should be carefully controlled to create a balance between the ammonia excreted into the medium and the ammonia converted to glutamine.

Photosynthetic activities of cells and organelles can be stablised by immobilisation on solid supports (Hall et al., 1987, Gisby et al., 1987). Here we report the results of our studies on ammonia and hydrogen production by *A.azollae* immobilized on polyvinyl and polyurethane foams under different experimental conditions. We also present the morphological features of cells immobilized on solid supports as revealed by low temperature scanning electron microscopy.

When TiO2 semiconductor electrodes coated with dyes (sensitizers) are illuminated, the dye molecules become excited and inject electrons into the conduction band of TiO2. If these electrons can be channelled to a counter electrode (cathode), a photocurrent can be generated. If instead of a dye, the TiO2 electrode is coated with PSI particles and then illuminated with white light, the excited P700 (PSI reaction centre chlorophyll) will inject electrons into TiO2. For continuous electron injection, a reductant, e.g. ascorbate, should be added to the electrolyte which would reduce P700+ back to P700.

$$P700 + TiO_2 \xrightarrow{h\nu} P700^+ + TiO_2 \text{ e cb}$$
$$P700^+ + \text{Ascorbate} \longrightarrow P700 + \text{dehydroascorbate}.$$

A coloured charge transfer complex formed by absorption of ferrocyanide at the surface of TiO2 particles and electrodes, on photo-excitation injects electrons into the conduction band of TiO2; the conduction band electrons can be used to generate a photocurrent (Gratzel, 1987).

$$Fe(CN)^{4-} + TiO_2 \xrightarrow{h\nu} Fe(CN)^{3-} + TiO_2 \text{ e cb}$$

For continuous electron flow, the ferricyanide should be reduced back to ferrocyanide. This can be achieved by coupling the electrode system with PSII particles which could photoreduce ferricyanide in the presence of a quinone; with the electron source being water.

We report the continuous generation of photocurrents from TiO2 photoelectrodes coated with PSI and PSII particles.

MATERIALS AND METHODS

Anabaena azollae, a presumptive isolate from *Azolla filiculoides* capable of growth in independent cultures (obtained from Dr.E.Tel-or, Rehovot, Israel) was grown at 32 C in a BG-11 medium without nitrate. Immobilization in polyvinyl and polyurethane foams supplied by Caligen Foam Ltd, Accrington, U.K., was performed according to Brouers and Hall (1986). For ammonia production experiments, small pieces of foam (5mm cubes) with immobilized cells were packed in a reactor consisting of a column (2cm internal diameter, 25cm high) containing 50ml nutrient medium. Fresh nutrient medium (BG11, with or without MSX) was added at the top of the

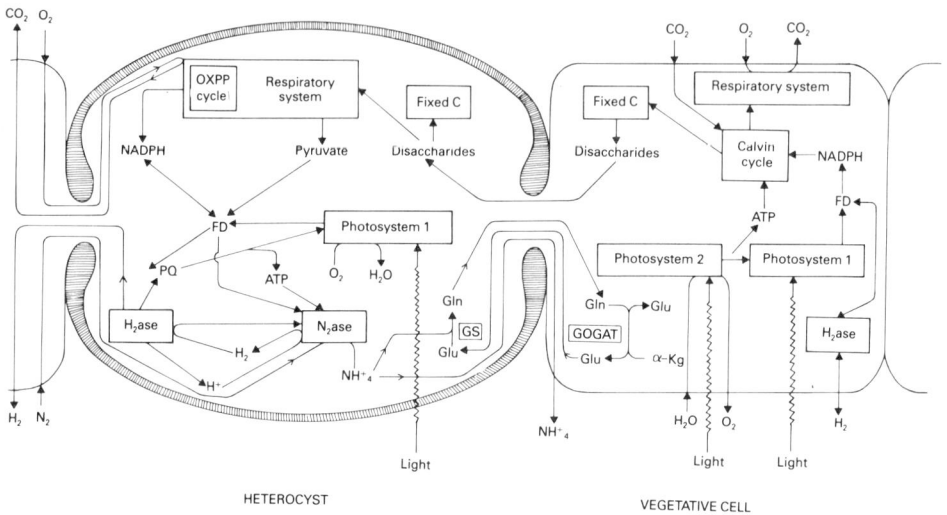

Fig. 1. Energy metabolism in heterocystous cyanobacteria (refer to Shi et al. for details)

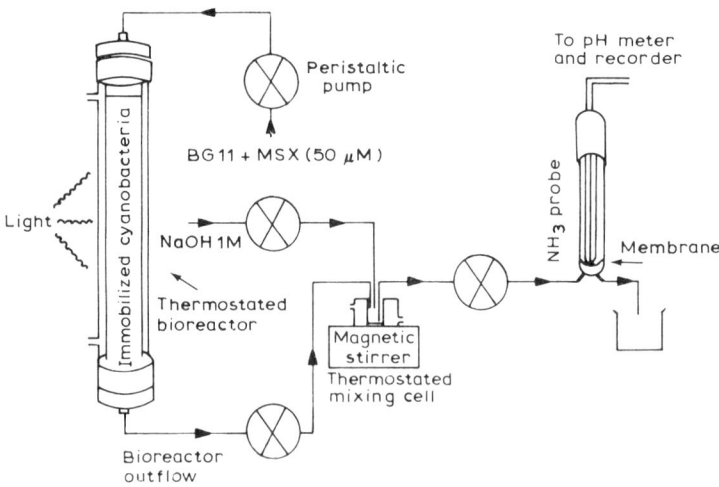

Fig. 2. Scheme of a bioreactor utilized for monitoring continuous ammonia production from polyvinyl foam-immobilised cyanobacteria (Vincenzini et al., 1986)

column and collected from the bottom using peristaltic pumps as shown in Fig. 2. The dilution rate was 0.4 hr -1 (flow 20ml h -1).

The outflow was passed through a mixing cell (5ml total volume) for continuous addition of 1M NaOH + 0.1M NaEDTA (10% v/v) in order to liberate NH_3. The resulting NH_3 solution was pumped to an ammonia electrode fitted with a flow device for continuous monitoring. Calibration was performed by pumping standard NH_4Cl solution through the mixing cell. Samples of the effluent were collected from time to time for parallel measurement of ammonia by the colorimetric method. The reactor column and mixing cell were thermostated at 28 C. Cool white fluorescent lamps were used for illumination of both sides of the biorector (intensity 100 umoles photons m -2 s -1 at the surface of the column). The methods for extraction and estimation of chlorophyll and ammonia were as described (Brouers and Hall, 1986).

HYPOL 2002, a special grade of polyether polyisocyanate capable of producing hydrophilic polyurethane foams, was obtained from W.R. Grace Ltd., London N.W.10, U.K. An aqueous suspension of A.azollae was mixed with HYPOL in equal proportions (w/w) at 0 C and the mixture subsequently warmed to 25 C to complete the polymerisation process. A blue-green foam was obtained in which the cyanobacterial cells were uniformly distributed. The foam was rinsed in the growth medium a few times (to remove any free polyisocyanate) and then cut either into small pieces or strips. The foam immobilized cells were incubated under the same growth conditions as the free-living cells. Nitrogenase activity as rates of acetylene reduction and H_2 production were assayed, heterocyst frequency determined and examination of the morphology of the immobilized and free-living cells by scanning electron microscopy (SEM) done, as described by Shi et al. (1987a). Photosystem I particles were prepared from spinach chloroplasts by a modification of the method of William Smith et al. (1978) and stored in liquid nitrogen. The preparation used had a chlorophyll concentration of 0.79mg per ml and an oxygen consumption rate of 2700/u moles per h. per mg Chl, determined in an oxygen electrode in the presence of 25mM ascorbate, 2.5mM TMPD and 50uM methyl viologen. Photosystem II particles were isolated from spinach and pea chloroplasts according to Ford and Evans (1983) and stored in liquid N_2 in a medium containing 20mM MES buffer pH 6.5, 5mM $MgCl_2$ and 20% glycerol. The spinach particles had an oxygen evolution activity of 173 umoles per mg. Chl.h and the pea particles 177 umoles of O_2 per mg. Chl.h when assayed in an oxygen electrode in the presence of 1mM ferricyanide and 1mM 2,6 dimethyl benzoquinone (DMBQ).

The semiconductor electrodes used had a surface area of 4 cm^2 and were coated with TiO_2 (anatase), TiO_2 + 10% Al, or TiO_2 + 5% Al + 3% Nb. Ferrocyanide coated electrodes were prepared by immersing the electrodes in a deaerated (argon flushed) solution containing 10mM perchloric acid and 10mM potassium ferrocyanide for one h. and then washing in distilled water and drying in air. The electrodes were coated with NAFION, where applicable, by layering 150 ul of 0.1% ethanolic solution of NAFION (a perfluorinated ion exchange powder, Aldrich Chemicals) on the surface of the electrode and drying in argon. For immobilization of photosystem particles on the electrode surface 5-10 ul of the particle suspension were uniformly layered on the electrode and then dried in a current of argon. When electrodes were reused after photocurrent measurements, they were washed in acetone to remove the chlorophyll followed by 1N NaOH, 1N HCl and distilled water.

Photocurrent measurements were performed by immersing the semiconductor electrode in an electrolyte containing 50mM NaCl or $CaCl_2$ buffered with 10 mM MES pH 6.2 or 10mM HEPES pH 7. The electrolyte, 150ml, was taken in a glass cell provided with a quartz window. The cathode was a platinum electrode and a calomel electrode introduced in the cell served as reference. The electrolyte was flushed throughout the experiment with argon to maintain anaerobicity and also to stir the electrolyte. The light source was a xenon lamp and a 430 nm cut off filter served to eliminate UV radiation falling on the electrodes: The electrode was connected to a galvanometer to record the current generated. All measurements were carried out at a light intensity of 500W m -2.

RESULTS AND DISCUSSION

Studies with immobilized cyanobacteria. The heterocyst frequency of Anabaena filiculoides immobilized in polymer foam pieces was double that of free-living cells and reached a maximum of 17% in 20 days; the frequency of heterocysts is 6-8% for free-living cells and 30-36% in symbiotic association with Azolla. In batch experiments (cultures grown in 250ml Erlenmeyer flasks) the nitrogenase and hydrogenase activities as well as ammonia photoproduction rates with MSX of the immobilized cells were higher than those of free-living cells (Shi et al., 1987a). An interesting observation was that the immobilized cells secreted ammonia, albeit in low quantities, even in the absence of MSX - in this respect the foam-bound cells mimic the behaviour of symbiotic cells. Low temperature S.E.M. of the immobilized cells revealed a thick mucilage layer covering the surface of both the cells and foam matrix; this phenomenon closely resembles the attachment of the symbiont Anabaena in the Azolla leaf cavity (Shi et al., 1987b) (Fig. 3).

Ammonia production in a biophotoreactor. Initial experiments showed the requirements for optimal continuous production of a flow-through reactor packed with foam immobilized cyanobacteria. Results indicated that at the dilution rate of 0.4 h -1 and in the presence of MSX, ammonia production ceased after about 70 h. The ability to produce NH3 could, however, be partly restored after a 24 h period in the absence of MSX. It was also shown that ammonia production rapidly ceased in the dark. These preliminary results indicated that continuous presence of MSX was a major limitation for long term ammonia production. In further experiments, twin packed bed reactors with immobilized A.azollae (one month after foam immobilization) were run in parallel under alternative pulses of MSX. Reactor 1 was operated under various dark-light cycles whereas reactor II was continuously illuminated for a period of 200 h before starting dark-light cycles. After an initial 10 h MSX treatment in both reactors, the production of ammonia in the effluent was monitored. Maintenance of a net ammonia production was then observed for 12 h following the removal of MSX, whereupon a plateau was reached with an ammonia concentration ca. 2.2 10 -4 M. A second MSX period resulted in a further increase in ammonia production. The maximal ammonia concentration in the effluent was 5.5 10-4 M; it then decreased slowly with stabilization at 0.4 10 -4 M, 130 h after exposure to the second MSX pulse.

When dark-light phases were applied (reactor I) net production of ammonia was observed during the light periods in the absence of MSX (following the first MSX, 10 h period), whereas the ammonia concentration decreased rapidly during the dark periods. The rate of ammonia production in the light remained stable for seven successive light-dark cycles without MSX addition, following the second 10 h MSX treatment (NH3 concentration in effluent at the end of the light period ranging from 2.0 to 3.0 10 -4 M); the ammonia concentration then decreased during the subsequent four cycles.
Further ammonia production could then be obtained by a new 10 h treatment with MSX resulting in subsequent maintenance of high ammonia production rates during the following dark-light cycles in MSX-free medium (Brouers

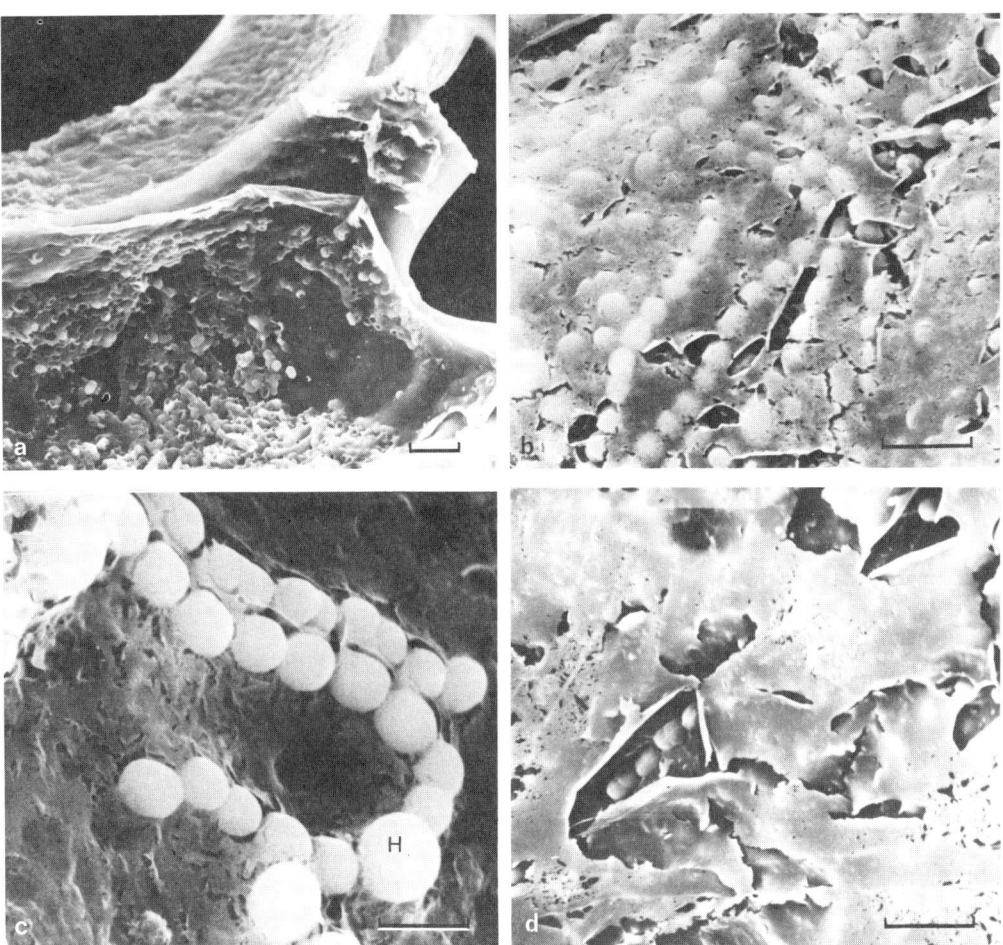

Fig.3. Scanning electron micrographs of frozen-hydrated A.azollae immobilized on polyvinyl foam PR22/60 (10 d old): a-c were lightly etched; d was sublimed for about 40 minutes. a. Low-power view showing densely packed filaments on surface of foam. Bar = 20 um. b. Filaments of cells underlying a thick layer of mucilage. Bar = 10 um; c. Medium-power view of filaments. Note high heterocyst (H) frequency. Bar = 5 um; d. Sublimed material showing characteristic pattern of splitting and folding in mucilage layer; note cells being revealed under film. Bar = 10 um.

and Hall, 1986; Brouers et al., 1987). Similar results were obtained in ammonia production studies using polyvinyl foam-immobilized Cyanospira rippkae cells in a continuous flow-packed bed photoreactor operating under dark-light cycles (Vincenzini et al., 1986).

HYPOL-immobilized A.azollae. After an initial decrease of cell contents, the foam pieces became densely populated with cells. It was found that the constant shaking of the conical flasks in which the cells and foams were incubated causes loss of cells at all exposed surfaces of the foam pieces, as they constantly rub against each other. A lower shaking frequency increased cell density on the foam pieces, but cannot be too low because of diffusion problems that then arise. This same effect was found with immobilisation on preformed polyvinyl (PR 22/60) foam.

In experiments with a packed bed photobioreactor, continuous ammonia production by immobilised A.azollae was maintained for over a month, by giving weekly MSX pulses. Adding MSX continuously was not successful. Although ammonia production was maintained for a long time in the packed bed reactor, no further cell growth was observed and in the end ammonia production ceased and cell lysis occurred. It was however possible to revive activity in foam pieces taken after six weeks from a packed bed column, by placing them in fresh growth medium in conical flasks, and shaking them in an orbital incubator (still giving weekly MSX-pulses). From the results of these experiments, it was concluded that diffusion limitation had been a major problem in the packed bed reactor, but also that the indiscriminate, regular MSX-pulsing had been too harsh.

As the packed bed reactor was found unsuitable for prolonged ammonia production, a fluidised bed reactor was used to obtain a continuous culture of immobilized A.azollae. Immobilization was started in the reactor. instead of foam pieces that already contained A.azollae, first sterilised polyvinyl foam pieces were added to the medium. They were found to circulate reasonably at random through the reactor. Then immobilization was initiated by adding a small amount of free living A.azollae. The reasoning behind this was, that if cell growth was possible in this reactor, then diffusion problems such as found in the packed bed column were not likely to occur and future problems would be more likely linked to the addition of MSX. It was found that immobilization occurred within hours, and that cell growth was very fast. Because of the continuous tumbling and collisions, however, cell loading of the foams remained low. Also, "foam loading" is relatively low in a fluidised bed column, and therefore ammonia production would only result in low concentrations of ammonia in the effluent of the column, for which reason this column design was also abandoned (de Jong and Hall, unpublished).

Other designs known from the literature all posed similar or their own particular problems. We have come up with an apparently novel solution. This new design was found to enable rapid immobilization (even faster than the fluidised bed reactor), very high cell loading, and good diffusion properties. Where in a packed bed column, highest production rates were in the order of 300 umol/l.h, in this column a production rate of 1,000 umol/l.h was found, and viability of cells is maintained. Different configurations are to be tried out in the future.

Anabaena azollae does not produce ammonia in the isolated state. Ammonia production is induced by addition of the glutamine synthetase (GS) inhibitor MSX. At present, in the experiments on ammonia production by Anabaena azollae in a packed bed column, we have not examined the time response and dependence of GS activity on different concentrations of the inhibitor. In vitro, GS inhibition by MSX is reversible at low concentrations and irreversible at higher concentrations. We have now performed experiments with free living cells, in which it was found that it can take from one to eight hours before GS activity irreversibly disappears (at high concentrations, 10-100 uM) of MSX added to the growth medium With immobilized cells, in an experiment with a packed bed column, addition of 25uM MSX to the growth medium for ten hours did not initiate ammonia production. Normally in column experiments, about 100 uM MSX was used to initiate ammonia production.

Photocurrent generation by semiconductor immobilized photosystem particles

The capacity of photosystem particles to bind to TiO2 particles and retain their photosynthetic activities was studied before immobilizing the particles on to the electrodes. Both PSI and PSII were able to bind TiO2 powder with retention of 80% PSI activity and 50% PSII activity.

Photosystem I. The quantity of photocurrent varied with the type and age of electrode as is evident from the data in Table I. Some freshly made TiO2 electrodes generated low currents even in the absence of PSI. The incident photon to current conversion efficiency, defined as the number of electrons injected by the excited sensitizer (and recorded as photocurrent) divided by the number of incident photons, was calculated from the equation:

$$n\ (\%) = \frac{1240 \times \text{photocurrent density (uA/cm 2)}}{\text{wavelength (nm)} \times \text{photon flux (W/m 2)}}$$

The efficiency values in Table I are calculated assuming an average wavelength of 550 nm for the white light. Although spinach PSI particles bound easily on freshly made TiO2 electrodes, when the electrodes were re-used, the binding was not efficient - the particles leaked out into the electrolyte. Thus, the "used" electrodes were coated with NAFION before depositing the photosystem particles.

Low potential electron relays accept electrons from excited P700 of PSI either directly or via X, the membrane-bound iron-sulphur centre. The effect of the following electron carriers on photocurrent generated by PSI-coated electrodes were examined.
1. Spirulina ferredoxin [Em = -420 mV],
2. Methyl viologen [Em = -440 mV] and
3. V740 (1,1'-trimethylene-4,4'-dimethyl-2,2'dipyridylbromide),
   Em = -740 mV.

Addition of ferredoxin, the biological PSI acceptor, caused a dramatic fall in photocurrent. This may be due to a proportion of electrons from P700* being diverted to ferredoxin, and hence decreasing the TiO2-mediated photocurrent. There was no change in the photocurrent pattern from the TiO2 electrode (no PSI deposition) on adding ferredoxin to the electrolyte containing ascorbate. Addition of the very low potential electron relay V740 to the electrolyte resulted in a gradual increase in photocurrent with time. There is a possibility that V740 itself sensitizes the TiO2 electrode. Methyl viologen addition did not produce any major change in photocurrent generation pattern. In all cases with PSI, the photocurrent generation continued at almost constant rate during the measuring time, which lasted for 15 minutes or more; the photocurrent production ceased when a 660nm cut off filter was inserted in the light path.

Photosystem II. Spinach and pea PSII particles coated on different TiO2 based electrodes were used for photocurrent measurements in the presence of PSII electron acceptor DMBQ. In all experiments, addition of DMBQ resulted in an increase in photocurrent which remained constant for long periods. In control experiments with no deposition of PSII on the electrodes, there was no change in the photocurrent pattern on addition of DMBQ. Addition of the PSII oxygen evolution inhibitor DCMU caused an immediate fall in photocurrent, suggesting that the electron transport to the TiO2 electrode is linked to water photolysis.

TABLE I. Photocurrent generation from PSI-coated electrodes

Expt.No.	Type of electrode	Photocurrent Microamps.	Measuring time-min.	Conversion efficiency %
1.	TiO2	240	11	0.27
2.	TiO2.Al	124	5	0.14
3.	TiO2.Al.NAF10N	145	5	0.16
4.a.	TiO2	210	6	0.24
b.	+ Ferredoxin	110	14	0.14
5.	TiO2.Al.Nb	240	9	0.27
6.a.	TiO2-NAF10N	114	6	0.15
b.	+ V740	151	12	0.19

All measurements were at a potential of +0.2V against SCE in an electrolyte containing 50mM CaCl2 in 10mM MES buffer pH 6.5 with 1mM ascorbate as electron donor. Ferredoxin concentration in the electrolyte in 4b was 1.33 uM and V740 concentration in 6b was 50 uM. Light intensity 500 W.M^{-2}.

TABLE II. Photocurrent generation from PSII-coated electrodes

Electrode type	PSII source	Photocurrent uA	Conversion efficiency %
1. TiO2-ferrocyanide	Pea	5	0.006
2. TiO2-Al-Nb	Pea	6	0.007
3. Ru-derivatised TiO2	Pea	45	0.05
4. TiO2-Al	Spinach	9	0.01
5. TiO2-Al-Nb	Spinach	12	0.014
6a. TiO2-ferrocyanide	Spinach	5	0.006
6b. + 1 uM DCMU	"	1.	
7. TiO2-Prussian blue	Spinach	45	0.05

Measurements were made at a potential of +0.2V against SCE in an electrolyte containing 50mM CaCl2 in 10mM MES pH 6.2. The photocurrent values are those recorded after addition of 0.2mM dimethylbenzoquinone (DMBQ) to the electrolyte.

The photocurrents generated from PSII coated TiO2 electrodes were of low magnitude, 5-15uA. But PSII coated on a ruthenium dye-derivatised TiO2 electrode generated photocurrents of the order of 700 uA; however this electrode was not stable at pH 6 and the material leaked out into the electrolyte within a few seconds.

DISCUSSION

The main goal of our ongoing research programme is to explore the possibility of using photosynthetic organisms and constituents isolated from these organisms in biocatalytic systems for the efficient conversion of solar energy to useful products such as fuels, fertilisers and "high value" biochemicals. We have previously shown that photosynthetic membranes and particles immobilized on to solid supports can be used for the photoproduction of H2, NADPH2 and H2O2 (Rao and Hall, 1984; De la Rosa et al., 1986). We have also demonstrated that bacterial hydrogenases immobilized on TiO2 particles can be used for light-induced H2 evolution (Cuendet et al., 1986). Ammonia is a nitrogen fertiliser which cyanobacteria can release along with a variety of organic compounds during different stages of their growth. Normally, cyanobacteria do not release ammonia during their growth although there are a few strains and mutants, recently reported, which can secrete ammonia (Subramanian and Shanmugasundaram, 1986; Latorre et al., 1986). _Azolla-Anabaena_ occurring naturally in rice fields in Asia is a very important natural nitrogen fertiliser, and it is known that the nitrogen fixation rates of _Anabaena_ in the _Azolla_ symbiotic state is very high. This knowledge prompted us to study ammonia production by this cyanobacterium in an environment which mimics the natural niche of the organism i.e. inside polymer foam cavities. The results are so far encouraging. There are still problems to be solved and better reactor designs to be ac hieved for the continuous production of ammonia. Also, the cells should be able to produce ammonia, preferably without the addition of the glutamine synthetase inhibitor.

The response of oxide semiconductor photoelectrodes (as anodes) to solar radiation can be enhanced by chemisorbed dyes. The dye must be selected such that its ground state red-ox potential lies within the band gap of the semiconductor while that of the excited state lies above the conduction band edge (see Gratzel, 1987). On photoexcitation, the excited dye molecule injects an electron into the conduction band of the n-type semiconductor. The dye molecule, now in an oxidised state, returns to the initial state by reaction with an electron donor present in the electrolyte. At the cathode, a reduction takes place. If the anode reaction is the reverse of the cathode one, then the cell is a current producing system, analogous to a solid state photovoltaic cell. The P700 species (of the PSI reaction centre) has a red-ox potential of +.48V in the ground state and approx. -1.0V in the excited state. At pH 6.5, the conduction band potential of TiO2 is about -0.5V so that electrons can be readily transferred from P700* to the conduction band of TiO2. Our preliminary studies were done with spinach PSI preparations. Possibly better results can be obtained by using PSI preparations from the thermophilic cyanobacterium, _Phormidium laminosum._ That _Phormidium_ PSI contains all the elctron transfer components present in plant PSI was established by the identification of signals corresponding to components A1 and A0 by ESR spectroscopy (Smith et al., 1987). The ESR spectra are shown in Fig. 4. A PSI complex with a molecular mass of 600 kDa has been isolated from _Synechococcus_ sp. (Boekema et al., 1987). The semiconductor oxide electrode coupled to PSII preparations offers a novel and inexpensive

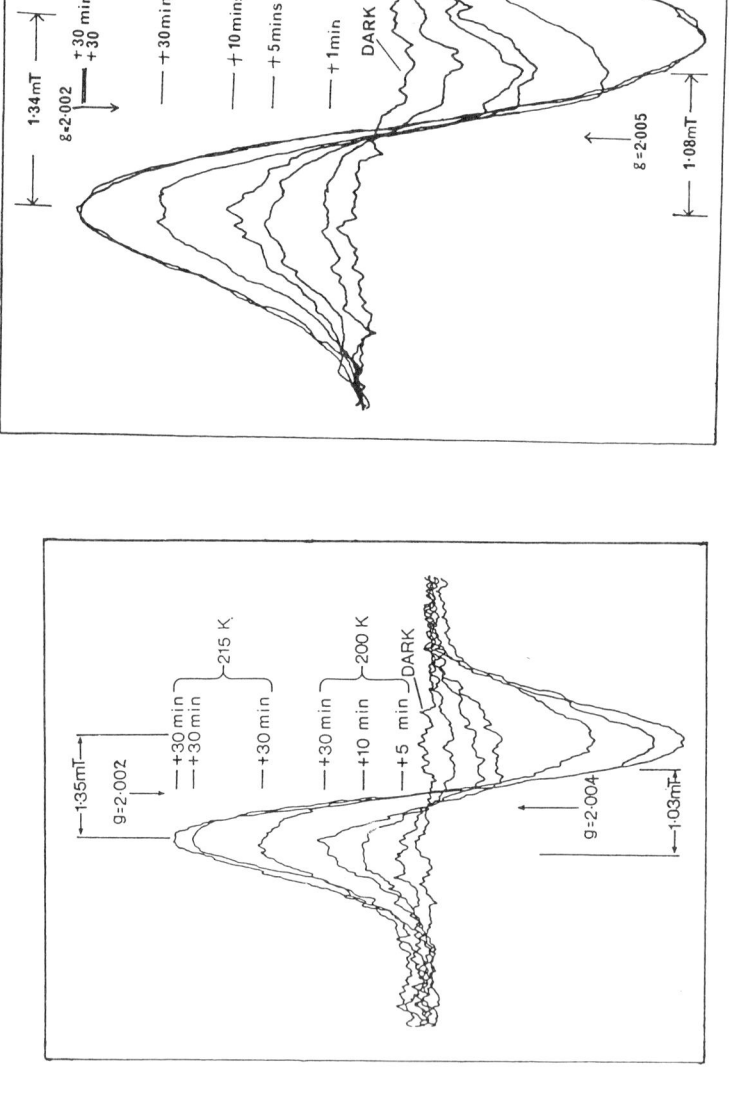

Fig. 4. ESR spectra of A_1 and A_0 of illuminated PSI particles. Left: Pea PSI at 200K (A_1) and 215K (A_1 and A_0). Right: Phormidium laminosum PSI at 205K and 230K (Smith et al., 1987)

route for the generation of photocurrent since the electron source in this system is water. Here again, much progress has still to be made to improve the efficiency of the system. The data in this report was obtained from PSII preparations from spinach and pea. Probably better photocurrent values could be obtained by using highly active preparations from Phormidium or Synechococcus sp. (Rogner et al., 1987).

ACKNOWLEDGEMENTS

We acknowledge with thanks the co-operation of M. Brouers and D-J. Shi in the ammonium production studies, J. Hubbard and L. Tilling in the preparation of photosystem particles and N. Vlachopoulos and P. Liska in the photocurrent measurements. DOH acknowledges financial assistance from the European Commission.

REFERENCES

Boekema, E.J., Dekker, J.P., van Heel, M.G., Rogner, M., Saenger, W., Witt, I. and Witt, H.T. 1987. Evidence for a trimeric organization of the photosystem I complex from the thermophilic cyanobacterium Synechococcus sp. FEBS Lett., 217, 283-286.

Brouers, M. and Hall, D.O. 1986. Ammonia and hydrogen production by immobilized cyanobacteria. J.Biotech., 3, 307-321.

Brouers, M., de Jong, H., Shi, D.J., Rao, K.K. and Hall, D.O. 1987. Sustained ammonia production by immobilized cyanobacteria. In "Progress in Photosynthesis Research", [Ed. J. Biggens) Vol.II Martinus Nijhoff Pub., Dordrecht. pp. 645-648.

Cuendet, P., Rao, K.K., Gratzel, M. and Hall, D.O. 1986. Light-induced H2 evolution in a hydrogenase-TiO2 particle system by direct electron transfer or via rhodium complexes. Biochimie, 68, 217-221.

De la Rosa, M.A., Rao, K.K. and Hall, D.O. 1986. Hydrogen peroxide phtoproduction by free and immobilized thylakoids. Photobiochem. Photobiophys., 11, 173-187.

Ford, R.C. and Evans, M.C.W. 1983. Isolation of a photosystem 2 preparation from higher plants with highly enriched oxygen evolution activity. FEBS Lett., 160, 159-164.

Gisby, P.E., Rao, K.K. and Hall, D.O. 1987. Entrapment techniques for chloroplasts, cyanobacteria, and hydrogenases. Methods in Enzymology, 135, 440-454.

Gratzel, M. 1987. Artificial Photosynthesis. Highly efficient sensitization of wide band gap semiconductors. In "Proc.of the 31st IUPAC Conf. Sofia, Bulgaria, July 1987 (In Press).

Hall, D.O., Brouers, M., de Jong, H., De la Rosa, M., Rao, K.K., Shi, D.J. and Yang, L.W. 1987. Immobilized photosynthetic systems for the production of fuels and chemicals. Photobiochem. Photobiophys. Suppl., 167-180.

Latorre, C., Lee, J.H., Spiller, H. and Shanmugan, K.T. 1986. Ammonium ion-excreting cyanobacterial mutant as a source of nitrogen for growth of rice: a feasibility study. Biotech. Lett., 8, 507-512.

Rao, K.K. and Hall, D.O. 1984. Photosynthetic production of fuels and chemicals in immobilized systems. Trends in Biotechnology, 2, 124-129.

Rogner, M., Dekker, J.P., Boekema, E.J. and Witt, H.T. 1987. Size, shape and mass of the oxygen evolving photosystem II complex from the thermophilic cyanobacterium Synechococcus sp. FEBS Lett., 219, 207-211.

Shi, D.J., Hall, D.O. and Tang, P.S. 1987a. Photosynthesis, Nitrogen fixation, Ammonia photoproduction and structure of Anabaena azollae immobilized in natural and artificial systems. In "Progress in Photosynthesis research" (Ed. J. Biggens), vol.II. Martinus Nijhoff Pubs., Dordrecht, pp.641-644.

Shi, D.J., Brouers, M., Hall, D.O. and Robins, R.J. 1987b. The effects of immobilization on the biochemical, physiological and morphological features of Anabaena azollae. Planta (in press).

Smith, N.S., Mansfield, R.W., Nugent, J.H.A. and Evans, M.C.W. 1987. Characterisation of electron acceptors Ao and A1 in cyanobacterial photosystem I. Biochim.Biophys.Acta, 892, 331-334.

Subramonian, G. and Shanmugaundaram, S. 1986. Uninduced ammonia release by the nitrogen fixing cyanobacterium Anabaena. FEMS Microbiology Letters, 37, 151-154.

Vincenzini, M., Brouers, M., Hall, D.O. and Materassi, R. 1986. Ammonia photoproduction by immobilized Cyanospira rippkae. Photobiochem. Photobiophys., 13, 85-94.

Williams-Smith, D.L., Heathcote, P., Sihra, C.K. and Evans, M.C.W. 1978. Quantitative EPR measurements of the electron-transfer components of the photosystem I reaction centre. Biochem.J., 170, 365-371.

TOWARDS THE DESIGN OF MOLECULAR PHOTOCHEMICAL DEVICES BASED ON RUTHENIUM BIPYRIDINE PHOTOSENSITIZER UNITS.

        Franco Scandola
        Dipartimento di Chimica dell'Università, Centro di Fotochimica
        CNR, 44100 Ferrara, Italy.

Abstract. Recent synthetic, spectroscopic and photopysical work related to the problem of light energy conversion is reported. In particular, new isocyano Ru(II) bipyridine photosensitizers are described, and a number of polynuclear complexes are discussed in which intramolecular transport of electronic charge and excitation energy can be studied.

1. Introduction.

  Living organisms possess molecular machineries capable of performing sophisticated photobiological functions (e.g., energy conversion, vision, etc.) [1]. These functions require a complex elaboration of the absorbed light energy input that is only possible through a suitable organization of molecular components in space. Obviously, there would be no hope in trying to synthetically reproduce the enormous complexity of natural systems. On the other hand, the design and synthesis of supramolecular structures made up of relatively small numbers of suitably assembled molecular components and capable of performing selected light-induced functions seems to be a non-unreasonable and stimulating goal. A supramolecular structure of this type can be called a molecular photochemical device. Looking towards this future goal, extended information on the peculiar photochemical and photophysical behavior of assemblies of molecular components (covalently linked bi- or polycrhomophoric sytems, donor-acceptor complexes, host-guest complexes, ion pairs, etc.) is needed. Research in this field is rapidly developing, establishing a new branch of photochemistry that can be conveniently called supramolecular photochemistry [2].
  Some of the possible functions of molecular photochemical devices have been classified [3] as follows:
- generation and migration of electronic energy;
- photoinduced vectorial transport of electric charge;
- photoinduced conformational changes;
- control of photochemical and photophysical properties of molecular components.
For each type of function, components to be assembled and possible types

of device have been discussed on general grounds [3]. Of the functions listed above, the second is certainly very relevant to the conversion of light into chemical energy, although functions of the first type can also be important in this context (see, e.g. energy collecting "antenna" devices to increase efficiencies).

In this report, we will describe some of our studies aimed at (i) obtaining new inorganic photosensitizers by second-sphere modification of known ones, and (ii) assembling photosensitizer units with other molecular components in discrete, covalently bound supramolecular structures. Studies of type (i), besides their intrinsic interest, have some relevance to the problem of how the properties of a photosensitizer are modified by inclusion in a supramolecular structure. Systems of type (ii) would be useful to study the basic processes of intramolecular electron and energy transfer involved in the performance of molecular photochemical devices.

2. New Photosensitizers: Isocyanide Bipyridine Mixed-ligand Ruthenium(II) complexes.

Most mixed-ligand 2,2'bipyridine Ruthenium(II) complexes do not exibit the outstanding excited-state properties of the parent $Ru(bpy)_3^{2+}$ complex, due to fast deactivation of the relevant Ru- bpy $d-\pi^*$ excited state via low-lying d-d states. This process can be avoided by the use of strong-field nonchromophoric ligands. For example $Ru(bpy)_2(CN)_2$ [4] and $Ru(bpy)_2(CN)_4^{2-}$ [5] have long-lived, emitting $d-\pi^*$ excited-states in fluid solution. These cyanide complexes, however, are extremely sensitive to the environment. In principle, analogous complexes with the stong-field isocyanide ligands could have long-lived excited states and be immune towards environmental effects. Thus, the methyl isocyanide complexes $Ru(bpy)_2(CN)(CNMe)^+$, $Ru(bpy)_2(CNMe)_2^{2+}$, $Ru(bpy)(CNMe)_4^{2+}$ have been synthetized and studied.

The complexes have been prepared by methylation of the corresponding cyanide complexes, followed by appropriate chromatographic separation procedures. The complexes are appreciably stable both thermally and photochemically. Due to the high electron-withdrawing effect of isocyanide, the substitution of cyanide by isocyanide ligands causes a progressive blue shift of the low-energy band in the absorption spectra. For the colorless $Ru(bpy)(CNMe)_4^{2+}$, the $d-\pi^*$ band is actually almost hidden by the $\pi-\pi^*$ bands of the bpy ligand.

As far as the emission is concerned, the important observation is that the 77 K emissions of $Ru(bpy)_2(CNMe)_2^{2+}$ and $Ru(bpy)(CNMe)_4^{2+}$ are completely different from those of the corresponding cyanide complexes indicating differences in the nature of the emitting states. The $Ru(bpy)(CNMe)_4^{2+}$ shows a pure $\pi-\pi^*$ phosphorescence. In the $Ru(bpy)_2(CN)_2$, $Ru(bpy)_2(CN)(CNMe)^+$, $Ru(bpy)_2(CNMe)_2^{2+}$ series a progressive mixing of $\pi-\pi^*$ character into a $d-\pi^*$ emission via configuration interaction was observed, so that for Ru(bpy)$_2$(CNMe)$_2^{2+}$ the emission appears to be largely but not yet completely π-π^* in character. A blue shift in emission, accompanied by a change in the nature of the emitting state has previously been observed for the cyanide complexes following protonation in highly acidic solutions[6,7].

At room-temperature, Ru(bpy)$_2$(CN)(CNMe)$^+$ exhibits a weak (Φ(H$_2$O) = 3 x 10^{-4}) d-π^* emission and Ru(bpy)$_2$(CNMe)$_2^{2+}$ does not emit appreciably. On the other hand, Ru(bpy)(CNMe)$_4^{2+}$ gives an intense (Φ(H$_2$O) = 0.013) long-lived (τ = 8.8μs) π-π^* emission (Fig. 1). Again, this behavior is

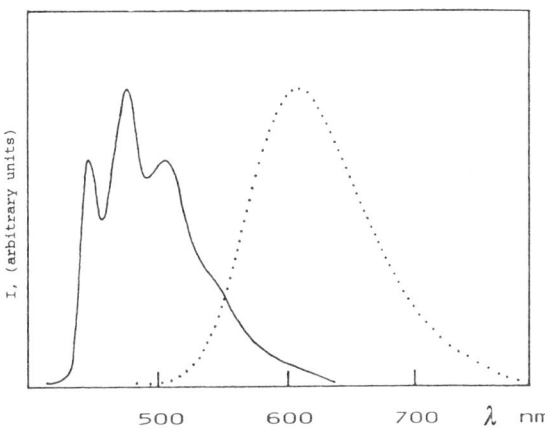

Fig. 1. Room temperature π-π^* emission of Ru(bpy)(CNMe)$_4^{2+}$ in aqueous solution (d-π^* emission of Ru(bpy)(CN)$_4^{2-}$ shown by comparison, dotted line).

reminiscent of that observed upon protonation [6,7] indicating that protonation and methylation cause very similar perturbation effects on the cyanide complexes.

Voltammetric measurements indicate that the substitution of cyanide by isocyanide ligands causes a large, progressive anodic shift in the oxidation potential which reflects the decrease in electronic density at the Ru center and is consistent with the observed spectral shifts. On the other hand, the reduction potentials of the isocyanide complexes are only slightly different from those of the cyanides, consistent with the fact that in the Ru(II) polypiridine complexes the reduction occurs at the polypyridine ligands. As the energy of the lowest excited state gradually increases upon methylation, a parallel increase in the oxidizing power of the excited state is expected. Such expectation has been verified for the Ru(bpy)(CNMe)$_4^{2+}$ complex by reductive quenching experiments.

Among the isocyanide complexes, Ru(bpy)(CNMe)$_4^{2+}$ stands out as a remarkable Ru(II) polypyridine photosensitizer in view of (i) its long lifetime in fluid solution, and (ii) its very high excited-state oxidizing power ($^*E_{1/2}$ ^{red} = 1.49 V vs SCE).

3. Cyano-bridged Supramolecular Systems Containing the Ru(bpy)$_2$$^{2+}$ Photosensitizer Unit.

Most of the interesting functions that molecular photochemical devices can perform are related to the occurrence of intramolecular (intercomponent) energy and/or electron transfer processes. Key components for this type of processes to be obtained are those that represent the interface of the device towards light, that may be called energy- or electron-transfer photo-sensitizers [3]. Common requisites for these two types of component are: (i) good light-absorbing properties, (ii) high excited-state energy, (iii) long excited-state lifetime, (iv) stability towards photochemical decomposition, (v) possibility of being linked, without loss of their useful characteristics, to other components. Specific requisites for electron transfer photosensitizers are: (vi) a useful partitioning of the excited-state energy between reductive power of the excited state and oxidizing power of the ground-state oxidized species (or vice-versa), and (vii) thermal inertness of the oxidized (or reduced) species.

Looking for inorganic energy- or electron-transfer photosensitizers to be used as components in molecular photochemical devices, the above requirements seem to be mostly (if not completely) met by the Ru(bpy)$_2$$^{2+}$ molecular fragment, provided that strong field ligands are used in the two vacant coordination sites to make the appropriate connections with other molecular subunits. Cyanide ligands appear to be a natural choice from this viewpoint. We have recently been studying the behavior of the Ru(bpy)$_2$(CN)$_2$ photosensitizer-bridge molecular fragment in a number of different supramolecular contexts, with particular attention to the routes for utilization (or degradation) of the excitation energy within the supramolecular structure. Depending on the metal-containing M subunits to which Ru(bpy)$_2$(CN)$_2$ is bound in the supramolecular structure, both intramolecular electron- and energy-transfer pathways have been found to be operative.

3.1. Intramolecular Electron Transfer.

Metal-containing subunits M capable of undergoing photoinduced intramolecular electron transfer (i.e., reduction or oxidation of M by the excited photosensitizer unit) should have redox potentials in the ranges $E_0(M/M^-) > -1.3V$ or $E_0(M^+/M) < +0.4V$ vs SCE, respectively (these values, taken from the redox properties of the excited free Ru(bpy)$_2$(CN)$_2$ (Table I), may change somewhat, as noticed before, upon metalation of the chromophore). The Ru(NH$_3$)$_5$$^{3+/2+}$ and Ru(NH$_3$)$_4$py$^{3+/2+}$

moieties should largely meet these requirements, judging from the redox potentials of related monomeric couples. The structures of a number of bi- and trinuclear species containing these subunits are schematically depicted in Fig. 2. For each structure, complexes corresponding to various combinations of oxidation states at the ruthenium centers (indicated by the numbers in square brackets) have been isolated and studied [8,9].

These polynuclear ruthenium complexes have very interesting charge-transfer absorption spectra, in which optical transitions can be identified that interconnect virtually all of the localized redox sites (both adjacent and remote) available in the supermolecule [8,9]. When the polynuclear complexes of Fig. 2 are excited into the Ru(bpy)$_2^{2+}$ chromophoric unit no d-π* emission is observed, indicating efficient intramolecular excited state quenching by the attached metal-containing units [8,9]. Although other mechanisms such as energy transfer or

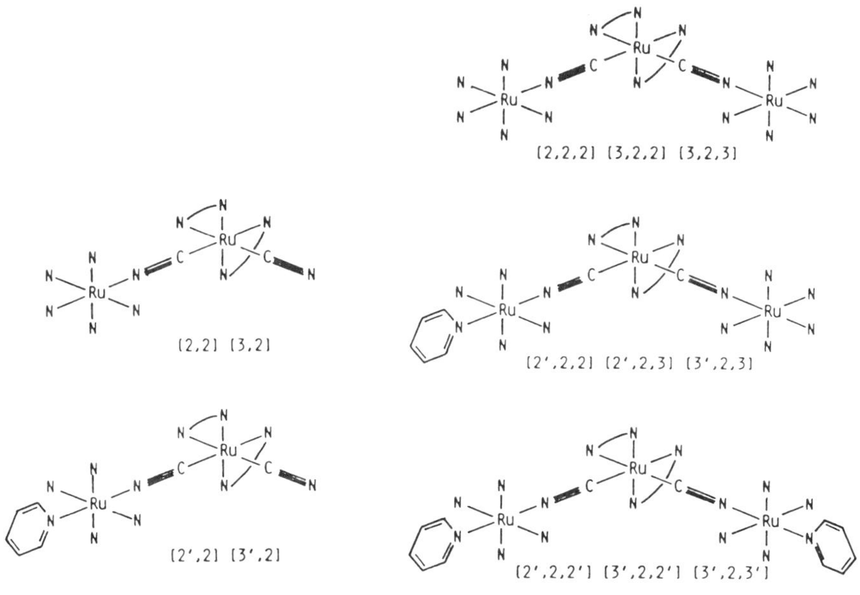

Fig. 2. Schematic structures of bi- and trinuclear complexes containing the Ru(NH$_3$)$_5^{3+,2+}$ and Ru(NH$_3$)$_4$py$^{3+,2+}$ subunits.

paramagnetic interactions cannot be ruled out in some of these systems, the most likely mechanism for intramolecular quenching is by far electron transfer. An example of the variety of electron-transfer deactivation pathways available in these multi-site systems is given in Fig. 3 for the py(NH$_3$)$_4$Ru-NC-Ru(bpy)$_2$-CN-Ru(NH$_3$)$_5$$^{5+}$ complex. The Figure points out, in one-electron terms, the relationship between the intermediates available for radiationless deactivation and the excited states responsible for the optical transitions in the absorption spectrum. The excited state responsible for the "remote" IT absorption (transition 5 in Figs 3) is conceptually similar to the charge-separated states that have been recently observed in porphyrin-based organic molecular "triads" [10,11]. No transient attributable to such

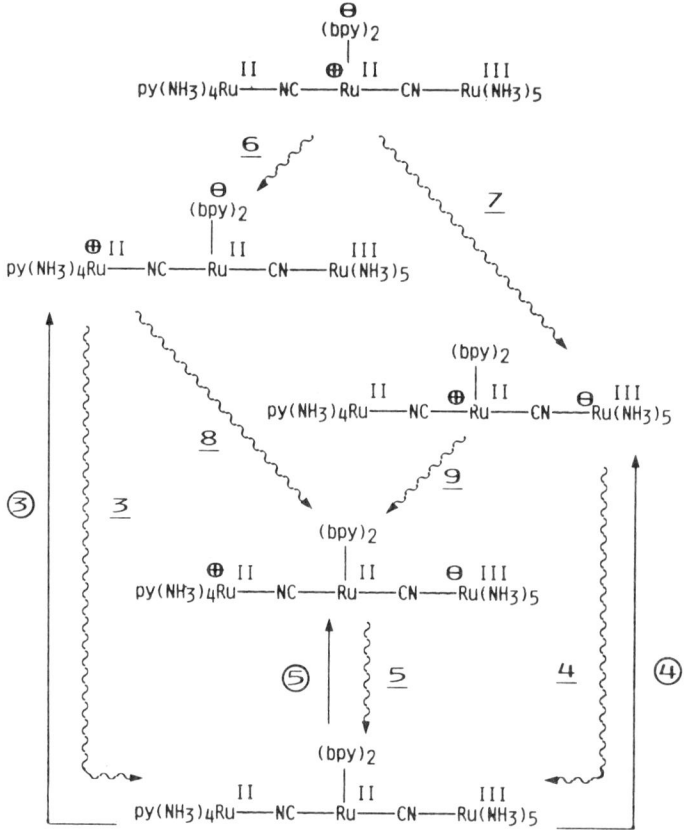

Fig. 3. Relationship between the electron transfer pathways for deactivation of the excited [2',2,3] complex and the optical charge transfer transitions.

charge-separated state is observed in these systems on a nanosecond timescale, indicating that this state is either populated very inefficiently or decays very rapidly. The first hypothesis is a likely one, since in the competition between the primary recombination process (e.g. 3, in Fig 3) and the secondary electron transfer step leading to the charge-separated state (e.g., 8 in Fig 3) the former is expected to prevail. In fact, both processes should have by symmetry the same H_{ab} (same degree of adiabaticity), but the primary recombination should have a lower barrier, being thermodynamically favoured over the secondary electron transfer process (both processes are presumably in the "normal" Marcus free energy region [12]).

It would appear that the possibilities of obtaining photoinduced charge separation with trinuclear systems containing the $Ru(bpy)_2^{2+}$ chromophoric unit are bound to the possibility of (i) using two bridges entailing different degrees of adiabaticity for primary recombination and secondary electron transfer (electronic control), or (ii) bringing the primary recombination in the inverted region while leaving the secondary electron transfer in the normal one (nuclear control).

3.2. Intramolecular Energy Transfer.

In order to observe intramolecular energy transfer from the excited $Ru(bpy)_2^{2+}$ chromophore to the attached M units, these units should be chosen so as to have low-lying ($E^{oo}(M/M^*) \lesssim 2$ eV) excited states. Moreover, it is important that the M units are such that redox reactions with the excited chromophore can be ruled out on thermodynamic grounds ($E^o(M^+/M) > +0.4V$ and $E^o(M/M^-) < -1.3V$ vs SCE), since exergonic intramolecular electron transfer quenching is expected to be very fast (see previous paragraph). Examples of systems containing such M units are described in this paragraph.

If the M unit is $Ru(bpy)_2CN^+$, dinuclear or trinuclear species (thereafter called "dimer" or "trimers") can be obtained [13]. The structures of such species (unknown conformation) are schematically shown in Fig. 4. The trimer structure shown in Fig. 4 is that of one of

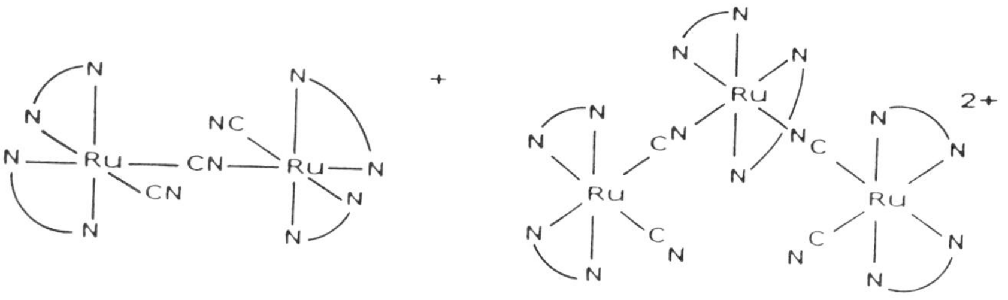

Fig. 4. Schematic structures of the bi- and trinuclear species ("dimer" and "trimer") obtained using the $Ru(bpy)_2CN^+$ subunit.

the three linkage isomers that are possible depending on the orientation of the two bridging cyanides (-NC-Ru-CN-, -NC-Ru-NC-, or -CN-Ru-NC-).

A combination of photophysical and redox data for these complexes [13] indicates that the emitting (lowest energy) units are the CN-Ru-CN one for the dimer and the CN-Ru-NC one for the trimer. For all these bi- and trinuclear species, the perfect agreement between the excitation and the absorption spectra indicates that intramolecular energy transfer from the higher energy chromophoric unit(s) to the lowest emitting one occurs with unitary efficiency. In the light of the results obtained with these bi- and trinuclear species, analogous polymeric species [14] should exhibit interesting energy conducting properties.

An example of the use of specific luminophoric M subunits to detect intramolecular energy transfer in a supramolecular system is given by the trinuclear complex shown in Fig. 5, where the $Ru(bpy)_2^{2+}$ chromophoric unit is bound via N-bonded cyanide groups to two $Cr(CN)_5^{2-}$ moieties [15]. In fact, $Cr(CN)_6^{3-}$ is known [16] to exhibit a very typical, sharp emission (Fig. 6) from the doublet ^2E_g state. In the trinuclear complex, the visible absorption spectrum is completely

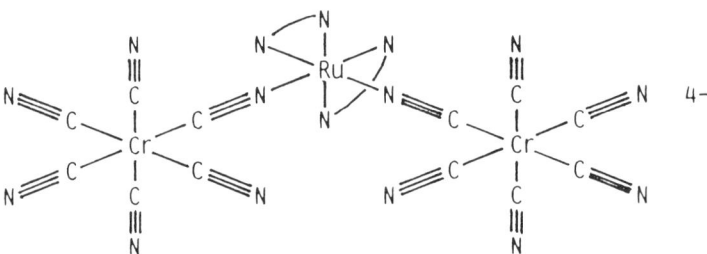

Fig. 5. Schematic structure of the trinuclear complex obtained using the $Cr(CN)_5^{2-}$ unit.

dominated by the d-π* transitions of the $Ru(bpy)_2^{2+}$ chromophore and yet the typical emission of the $Cr(CN)_6^{3-}$ moieties is obtained upon visible excitation (Fig. 6). Quantitative data indicate that the efficiency of conversion from the absorbing state of the Ru-based chromophore to the emitting state of the Cr-based luminophore is essentially unity [15]. This efficiency is higher than that of conversion from the light absorbing state to the emitting state within the isolated luminophore (0.5 [17]).

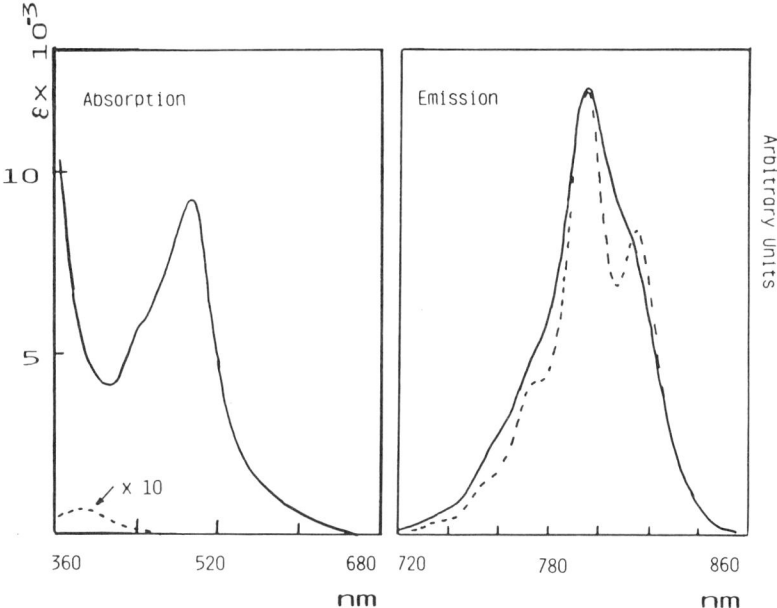

Fig. 6. Absorption (left) and emission (right) spectra of Ru(bpy)$_2$[Cr(CN)$_6$]$_2^{4-}$ (full line) and Cr(CN)$_6^{3-}$ (dashed line) in DMF.

This trinuclear complex is an example of how coupling to a photosensitizer may improve the performance of a luminophore by (i) increasing the cross-section for light absorption, (ii) achieving visible spectral sensitization, (iii) bypassing the inefficient intersystem crossing step of the luminophore. Light concentration (antenna) effects would be easily predictable for similar systems having the Cr(CN)$_6^{3-}$ luminophore bound to oligomeric or polymeric species of the type described above.

4. Conclusions

Based on their easily tunable photophysical and redox properties, transition metal complexes are versatile components to be used in the construction of photochemical molecular devices. The studies presented in this article show that the combination of the Ru(bpy)$_2^{2+}$ photosensitizer and cyanide bridging units allows the synthesis of a variety of polynuclear systems that exhibit interesting photochemical properties. Depending on the nature of the attached metal-containing units, supramolecular systems can be obtained that undergo efficient photoinduced intramolecular energy or electron transfer processes.

Acknowledgments. Drs. C. A. Bignozzi and M. T. Indelli have contributed much in terms of skillful work and good judgment to the studies discussed in this report. The work has been supported by the Consiglio Nazionale delle Ricerche and by the Ministero della Pubblica Istruzione.

REFERENCES
[1] See, e. g.,: Sauer, K in "Encyclopedia of Plant Physiology", Vol. 19, Staehelin, L. A.; Arntzen, C. J. (Eds.), Springer-Verlag, Berlin, 1986, p. 85.
[2] Balzani, V. (Ed.) "Supramolecular Photochemistry", Reidel, Dordrecht, 1987.
[3] Balzani, V.; Moggi, L; Scandola, F in ref. 2, p. 1.
[4] Demas, J. N.; Addington, J. W.; Peterson, S.H., Harris, E.W. J. Phys. Chem. 1977, 81, 1039.
[5] Bignozzi, C.A.; Chiorboli, C.; Indelli, M.T.; Rampi Scandola, M.A., Varani, G.; Scandola, F. J. Am. Chem. Soc. 1986, 108, 7872.
[6] Peterson, S.H.; Demas, J. N. J. Am. Chem. Soc. 1979, 101, 6571.
[7] Indelli, M.T.; Bignozzi, C.A.; Marconi, A.; Scandola, F. Proceedings of VII ISPPCC, Springer-Verlag Berlin, 1987.
[8] Bignozzi, C. A.; Roffia, S.; Scandola, F. J. Am. Chem. Soc. 1985, 107, 1644.
[9] Bignozzi, C. A.; Paradisi, C.; Roffia, S.; Scandola, F. Inorg. Chem., in press.
[10] Wasielewski, M. R.; Niemczyk, M. P.; Svec, W. A.; Pewitt, E. B. J. Am. Chem. Soc. 1985, 107, 5562.
[11] Gust, D.; Moore, T. A.; Liddell, P. A.; Nemeth, L. R.; Makings, L. R.; Moore, A. L.; Barrett, D.; Pessiky, P. J.; Bensasson, R. V.; Rougée, M.; Chachaty, C.; De Schryver, F. C.; Van der Auweraer, M.; Holzwarth, A. R.; Connolly, J. S. J. Am. Chem. Soc. 1987, 109, 846.
[12] Marcus, R. A.; Sutin, N. Biochim. Biophys. Acta 1985, 811, 265.
[13] Bignozzi, C. A.; Roffia, S.; Chiorboli, C.; Davila, J.; Indelli, M. T.; Scandola, F. manuscript in preparation.
[14] Bignozzi, C. A.; Scandola, F. Inorg. Chim. Acta, 1984, 86, 133.
[15] Bignozzi, C. A.; Scandola, F. manuscript in preparation.
[16] Wasgestian, H. F., J. Phys. Chem. 1972, 76, 1947.
[17] Sabbatini, N.; Scandola, M. A.; Balzani, V. J. Phys. Chem. 1974, 78, 541.

PHOTOINDUCED CHARGE-SEPARATION IN MODELS FOR PHOTOSYNTHESIS

J. W. Verhoeven*, H. Oevering*, M.N. Paddon-Row#, J. Kroon*, A.G.M. Kunst*

* University of Amsterdam,
Laboratory of Organic Chemistry and Laboratory of Physical Chemistry,
Nieuwe Achtergracht 129 and 127, 1018 WS Amsterdam, The Netherlands
\# University of New South Wales, Department of Chemistry,
Kensington NSW 2033, Australia.

ABSTRACT
 A series of molecules is described containing a π-electron donor moiety and a π-electron acceptor moiety interconnected by rigid, saturated hydrocarbon bridges. The length of the bridges is varied from 4 to 12 carbon-carbon bonds to provide effective edge-to-edge distances between donor and acceptor that range from 4.6 to 13.5 Å.
Upon photoexcitation fast intramolecular charge-separation is observed similar in rate and in temperature dependence to the early steps of photosynthesis. It is proposed that the bridges play an active role in mediating electron transfer via through-bond interaction, thereby enhancing the rate of charge separation significantly as compared to other model systems.
The consequences of these findings for the mechanisms underlying the very fast charge separation processes occurring in the early stages of photosynthesis are discussed.

INTRODUCTION

One of the most fundamental chemical changes brought about by the absorption of light is the promotion of electron transfer between molecular entities, conveniently designated as D(onor) and A(cceptor):

$$D + A \xrightarrow{h\nu} D^{+\cdot} A^{-\cdot}$$

Natural photosynthesis provides the most dramatic demonstration of the potential hidden in this basic photoreaction. In (bacterial) photosynthesis a chlorophyll-dimer $(BC)_2$ - the "special pair" - receives the radiation energy and thereby gains the energy required to enable it to transfer an electron to a pheophytin moiety (BP), an act occurring within 2-3 picoseconds (Martin et al. 1986) even at very low temperatures. Subsequently the electron is transferred to a quinone acceptor (MQ), which once again occurs (Holten et al. 1978) on a very short time scale of about 230 ps.

Crystallization in a form suitable for X-ray structure analysis of the protein complex containing this part of the photosynthetic unit has resisted all efforts until Deisenhofer et al. solved such a structure for a complex isolated from *rhodopseudomonas viridis* (Deisenhofer et al., 1984). In this complex the special pair $((BC)_2)$ and the pheophytin (BP) were found to be spatially separated by a centre-to-centre distance of no less than ~17 Å, while the centre-to-centre distance between the pheophytin and the menaquinone (MQ) is ~14.3 Å (see Fig.1).

This implies that very fast electron-transfer occurs between π-systems that are separated by edge-to-edge distances in the order of 8-10 Å, which seems quite amazing in view of the extremely minor overlap expected to occur between π-wavefunctions over such distances!

Until recently very little quantitative experimental data concerning the distance dependence of electron–transfer rates were available. From experiments on electron transfer between statistically distributed donor and acceptor species in an inert glassy matrix it had been concluded (Miller et al., 1984) that the rate falls of sharply with increasing donor-acceptor separation and that at a given edge-to-edge separation R_e (in Å) the fastest rate (k in s^{-1}) achievable under optimally exothermic conditions would be given by the exponential expression eq. (1) :

$$k = 10^{13.9} \exp(-1.20 R_e) \qquad (1)$$

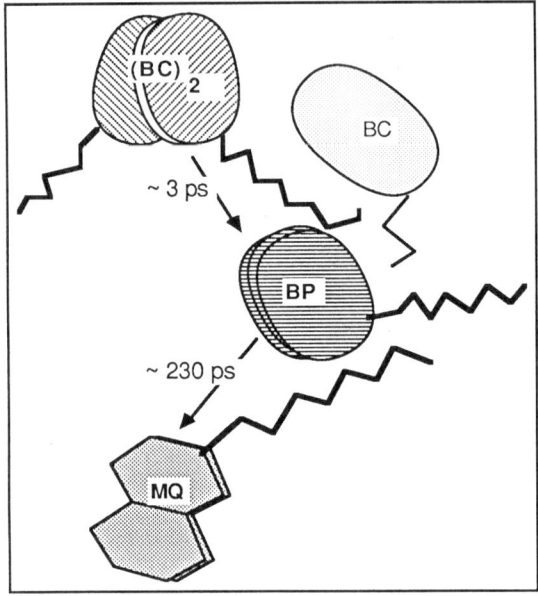

Fig. 1 Schematic representation of the spatial arrangement of the chromophores involved in the first steps of photosynthetic charge-separation in *rhodopseudomonas viridis*. Orientation of phytyl side-chains is indicated by wavy lines, furthermore the position of an additional bacteriochlorophyll unit is sketched.

From this expression shortest transfer times ranging from ~200 to ~2000 ps are predicted for edge-to-edge separations between 8 and 10 Å, i.e. two to three orders of magnitude longer than actually observed in the photosynthetic unit. This discrepancy seems to be aggravated by the results of a series of recent studies on electron transfer between redox sites attached (semi)synthetically to various molecular protein frameworks at crystallographically well defined locations. Thus McLendon et al. recently studied a series of (modified) cytochrome–c/cytochrome-b_5 complexes in which two heme-type redox sites are separated by an edge-to-edge distance of ~8.5 Å (16 Å centre-to-centre) (McLendon and Miller, 1985). One of the redox sites was modified to achieve a situation of optimal exothermicity which led to a transfer time of 2 microseconds (2×10^6 ps), i.e. about a million fold slower than the transfer occurring in the primary steps of photosynthesis under conditions that appear to be quite similar geometrically !

We now report that we have observed intramolecular photoinduced charge separation similar to that in the primary steps of photosynthesis both with respect to the distances involved and to the transfer times required in relatively simple model systems (see Fig. 2) where a rigid, saturated hydrocarbon spacer keeps the donor and acceptor sites at a well defined distance and orientation. Comparison of the rates achieved in these model systems with literature data on charge migration in other bichromophoric systems as well as other evidence leads to the conclusion that the saturated hydrocarbon spacers must play an active role in mediating the very fast electron transfer observed. The possible consequences of these observations for the mechanism of charge-separation in the primary steps of photosynthesis are discussed.

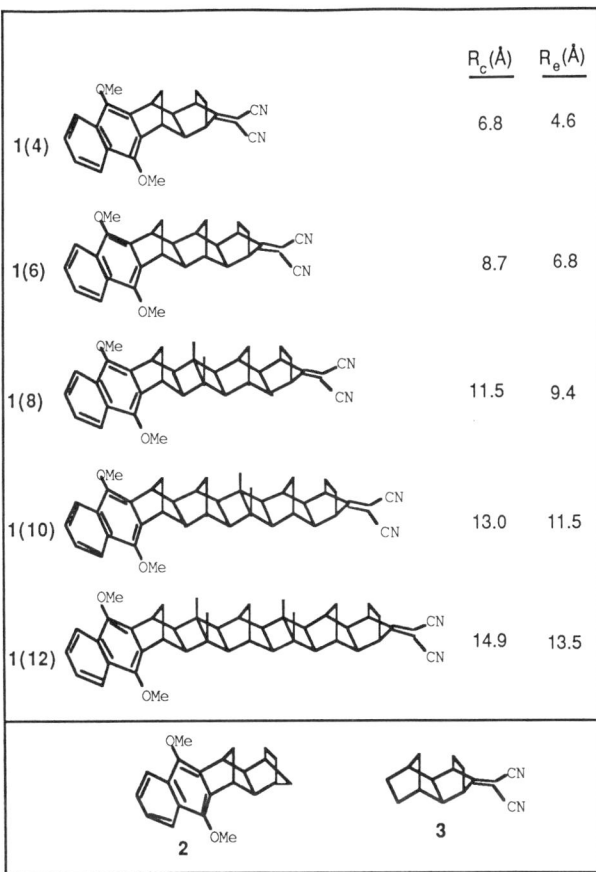

Fig. 2 Structure of the model systems studied

RESULTS AND DISCUSSION

Figure 2 shows the structures of the bifunctional model systems **1(n)**, (n = 4,6,8,10,12), investigated. The synthesis of these systems and a discussion of their electrochemical properties as well as of their electronic absorption and emission spectra have been given elsewhere (Oevering et al., 1987). These molecules contain a 1,4-dimethoxynaphthalene unit as the photoexcitable electron donor ($E_{1/2} = +1.1$ V vs. sce in acetonitrile) and a 1,1-dicyano- vinyl moiety as a moderately powerful electron acceptor ($E_{1/2} = -1.7$ V) separated by an array of at least **n** C–C sigma-bonds. The centre-to-centre separation (R_c) and the edge-to-edge separation (R_e) of the chromophores is also indicated in Fig.2. The latter is defined as the shortest atom-to-atom distance between the donor- and acceptor π-systems, whereas the former refers to the distance between the centre of the naphthalene unit and the midpoint of the exocyclic C=C bond of the acceptor. X-Ray structure data of **1(6)** and of precursors of **1(8)** and **1(10)** were used to evaluate these distances (Craig and Paddon-Row, 1987).

As reported earlier (Oevering et al., 1987; Warman et al., 1986) intramolecular electron transfer following photoexcitation occurs for all values of **n** and in a variety of solvents. This electron transfer results in quenching of the typical dimethoxynaphthalene fluorescence for **1(n)** as compared to that for the "isolated donor" **2**. Determination of the rate constant (k) of this photoinduced charge-separation was achieved in a variety of solvents (Oevering et al., 1987) by comparison of the lifetime of the residual donor fluorescence in **1(n)** for n = 8,10,12 with that of the reference system **2** via eq.(2) :

$$k = 1/\tau - 1/\tau_{ref} \qquad (2)$$

Until recently we were unable to determine k for **1(4)** and **1(6)** via this method, since this not only requires a time resolution better than ~10 ps, but especially since the quenching of the donor fluorescence, that accompanies the electron transfer, makes the measurements extremely sensitive to the presence of minor, fluorescent impurities. After careful recrystallization a sample of **1(6)** was now obtained for which the level of impurity fluorescence is sufficiently low to detect the very short lived (~ 3-4 ps) residual donor fluorescence . A typical fluorescence decay as observed for this sample in ethylacetate via picosecond time correlated single photon counting (Bebelaar, 1986) is shown in Fig. 3. Via a biexponential reconvolution procedure the lifetime of the short component was determined to be 3.5±0.5 ps, while that of the impurity background is comparable to the lifetime of the isolated donor (~4500 ps) and thus probably stems from one or more species lacking the acceptor chromophore. Similar results were obtained in tetrahydrofuran (3±1 ps) and in acetonitrile (4±1ps) .

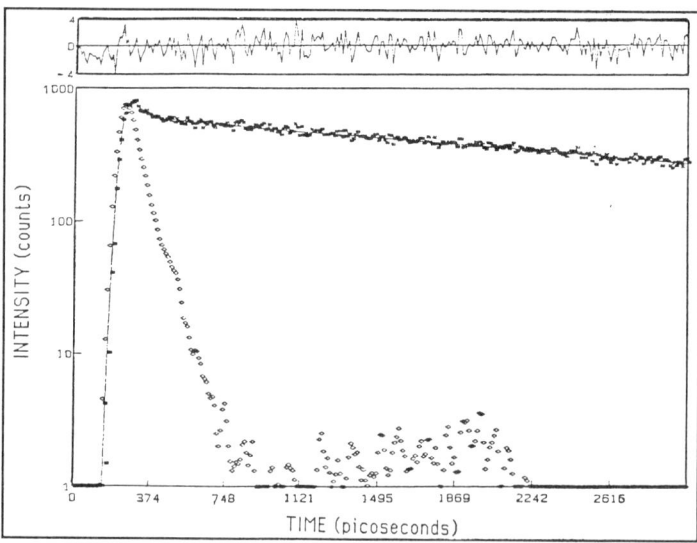

Fig. 3 Time resolved fluorescence (excitation at 303 nm, observation at 360 nm) for **1(6)** in ethyl acetate.
Filled dots indicate the experimental data, the continuous line gives the best fit. Open dots represent the laserpulse.

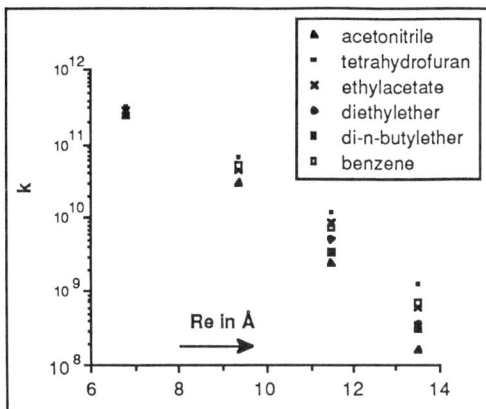

Fig. 4 Logarithmic plot of the rate of intramolecular photoinduced charge separation in **1(n)** (n=6,8,10,12) as a function of the edge-to-edge donor/acceptor separation (R_e) in various solvents.

In Fig. 4 a graphical representation is given of all data now available for the rates of intramolecular photoinduced charge-separation in **1(n)**,(n= 6-12), in a variety of solvents. From this plot it is evident that - as expected - a general exponential dependence of the rate on the distance exists and that the solvent dependence of the rate is relatively minor although in several cases the range of solvent polarities investigated extends from hydrocarbons to acetonitrile. As we have shown before (Oevering et al., 1987; Pasman et al., 1985) such minor solvent dependence is typical for a nearly barrierless electron transfer that occurs under conditions of optimal exo-thermicity. This is a very important observation, since it explains why the hydrophobicity of the environment, in which the first steps of photo-synthesis occur, has no detrimental effect on the rate of these processes.

The distance dependence of the electron transfer rate was investigated quantitatively in the two solvents of intermediate polarity in which k values for n = 4-12 are available.

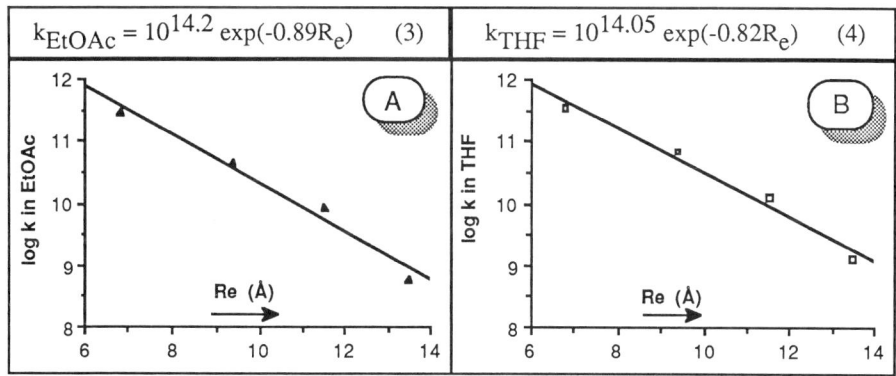

$$k_{EtOAc} = 10^{14.2} \exp(-0.89 R_e) \quad (3)$$

$$k_{THF} = 10^{14.05} \exp(-0.82 R_e) \quad (4)$$

Fig. 5 Distance dependence of the rate of photoinduced charge separation in **1(n)** in ethylacetate (Fig. 5A) and in tetrahydrofuran (Fig. 5B) as a solvent.

As shown in Fig. 5A and 5B an acceptable fit to a single exponential dependence of k as a function of R_e is found. From least squares regression analysis the expressions eq.(3) and eq.(4) were found to describe the distance dependence in ethylacetate and in tetrahydrofuran respectively, in both cases with a correlation coëfficient of r = 0.99.

While the preexponential factor in these equations is close to that in eq.(1), a much slower exponential decrease of k with increasing distance is found. Thus for the edge-to-edge distances occurring in the primary steps of photosynthesis (8-10 Å) eq.(4) predicts charge separation times of 6.2 - 32 ps, i.e. quite similar to those actually observed (see Fig. 1) in *rhodopseudomonas viridis*.

If one compares the present data for **1(n)** with those reported for various other systems such as intermolecular electron transfer in a glassy matrix (Miller et al., 1984), and electron transfer in protein complexes (McLendon et al., 1985; Isied et al., 1985) it seems obvious that one cannot hope to find a single distance dependence expression to encompass all these data. Apparently not only the structure of donor and acceptor but in particular also the nature of the medium interposed between these plays a decisive role. As we have argued before (Oevering et al. 1987; Krijnen et al., 1987) an array of saturated hydrocarbon bonds may act as an efficient mediator for electron transfer since through-bond interaction via such an array can provide significant electronic coupling over much larger distances than direct (through space) overlap (Hoffmann et al., 1968). For efficient through-bond interaction an extended all-trans conformation of the array of C-C sigma bonds is required, whereas attachement of this array to donor and acceptor should occur at a site of high local density in the relevant MO's. Both requirements are fulfilled in **1(n)**, which probably explains why the coupling in these systems seems to be more effective than in several other linked donor-acceptor systems containing different types of hydrocarbon spacers (Joran et al., 1984; Heitele et al.,1985; Closs et al., 1986; Gust et al., 1987)

Returning to the photosynthetic system it now seems plausible that the explanation for the short charge-separation times in the primary steps must be found in the nature of the medium between the redox-sites involved. In this connection it is interesting to note that saturated hydrocarbon chains (i.e. phytyl sidechains) extend from the special pair and from the menaquinone towards the intermediate bacteriopheophytin (see Fig.1). At this moment it is not clear, however, whether in *rhodopseudomonas viridis* any of these phytyl sidechains plays the role of a "molecular wire" (see also : Kuhn, 1986) that we attribute to the hydrocarbon bridges in **1(n)**. For *rhodopseudomonas sphaeroides* a fivefold decrease in the rate of the reverse electron transfer from the quinone (ubiquinone) to the bacteriopheophytin was recently reported to result upon removal of the isoprenoid sidechain from the quinone (Gunner et al., 1986).

Although no evidence has been found that the monomeric bacteriochlorophyll unit, which is intercalated between the special pair and the pheophytin (see Fig. 1), actually acquires an electron, it should be noted that also this unit may act as a mediator via a super-exchange mechanism (Kuznetsov and Ulstrup, 1982; Miller and Beitz, 1981), which is conceptually related to through-bond interaction.

Having shown that it is possible to construct simple model systems that mimic the rapid photoinduced charge separation of photosynthesis at ambient temperature, it appeared of interest to see whether these systems also share the remarkable temperature independence of the primary photosynthetic electron transfers (Bixson and Jortner, 1986; Kirmaier et al.,1985). Preliminary experiments do indeed suggest that this is the case, at least for **1(n)**, **n** ≤ 8. Thus for **n** = 4 and **n** = 6 virtually complete quenching of the donor fluorescence as a result of fast electron transfer persists even if solutions are frozen at liquid nitrogen temperature. For **1(8)** some preliminary measurements of the rate of charge-separation at various temperatures in apolar media are shown in Fig. 6. Once again no indications are found that the charge separation process slows down at lower temperatures, in fact a slight acceleration is observed in certain temperature regions, a behaviour also observed in photosynthesis.

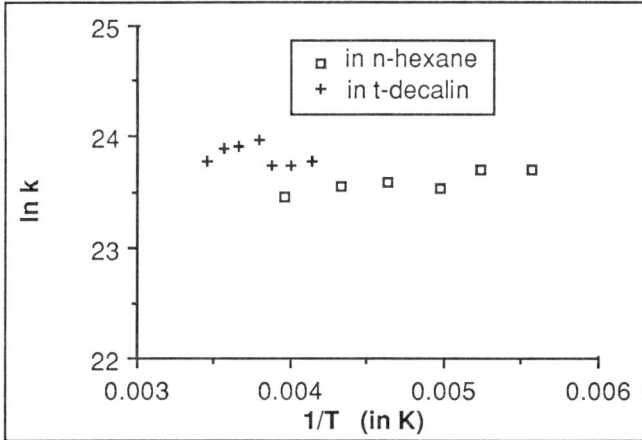

Fig. 6 Arrhenius plots for the rate of photo-induced charge separation in **1(8)** in two solvents

Thus these preliminary studies on the temperature dependence support our earlier proposal (Oevering et al., 1987) that in the lower homologues of **1(n)** photoinduced charge-separation occurs as a virtually barrierless process.

In conclusion the molecules **1(n)** consitute a series of model systems that not only provides donor-acceptor separations similar to those found in the photosynthetic unit, but also mimics the high rate and the temperature independence of the primary steps in photosynthesis.

ACKNOWLEDGMENT

We thank Ing. D. Bebelaar for his valuable contribution to the technical realization of the time correlated single-photon counting measurements. The present research was supported by the Australian Research Grants Scheme and by the Netherlands Foundation for Chemical Research (SON) with financial support from the Netherlands Organization for the Advancement of Pure Research.

REFERENCES

Bebelaar, D. 1986. *Rev. Sci. Instrum.*, 57, 1116-1125.
Bixson, M. and Jortner, J. 1986. *J. Phys. Chem.*, 90, 3795-3800.
Closs, G.L., Calcaterra, L.T., Green, N.J., Penfield, K.W. and Miller, J.R. 1986. *J. Phys. Chem.*, 90, 3673-3683.
Craig, D.C. and Paddon-Row, M.N. 1987. *Aust. J. Chem.* (in press).
Deisenhofer, J., Epp, O., Miki, K., Huber, R. and Michel, H. 1984. *J. Mol. Biol.*, 180, 385-398.
Gunner, M.R., Robertson, D.E. and Dutton, P.L. 1986. *J. Phys. Chem.*, 90, 3783-3795.
Gust, D., Moore, T.A., Liddell, P.A., Nemeth, G.A., Makings, L.R., Moore, A.L., Barrett, D., Pessiki, P.J., Bensasson, R.V., Rougée, M., Chachaty, C., De Schryver, F.C., Van de Auweraer, M., Holzwarth, A.R. and Conolly, J.S. 1987. *J. Am. Chem. Soc.*, 109, 846-856.
Heitele, H. and Michel-Beyerle, M.E. 1985. *J. Am. Chem. Soc.*, 107, 8286-8288.
Hoffmann, R., Imamura, A. and Hehre, W.J. 1968. *J. Am. Chem. Soc.*, 90, 1499-1509.
Holten, D., Windsor, M.W., Parson, W.W. and Thornber, J.P. 1978. *Biochim. Biophys. Acta*, 501, 112-126.
Isied, S.S., Vasillian, A., Magnuson, R.H. and Schwarz, H.A. 1985. *J. Am. Chem. Soc.*, 107, 7432-7438.
Joran, A.D., Leland, A., Geller, G.G., Hopfield, J.J. and Dervan, P.B. 1984. *J. Am. Chem. Soc.*, 106, 6090-6092.
Kirmaier, C., Holten, D. and Parson, W.W. 1985. *Biochim. Biophys. Acta*, 810, 33.
Krijnen, B., Beverloo, H.B. and Verhoeven, J.W. 1987. *Recl. Trav. Chim. Pays-Bas*, 106, 135-136.
Kuhn, H. 1986. *Phys. Rev. A*, 34, 3409-3425.
Kuznetsov, A.M. and Ulstrup, J. 1981. *J. Chem. Phys.*, 75, 2047.
Martin, J.L., Breton, J., Hoff, A.J., Migus, A. and Antonetti, A. 1986. *Proc. Natl. Acad. Sci. U.S.A.*, 83, 957.
McLendon, G. and Miller, J.R. 1985. *J. Am. Chem. Soc.*, 107, 7811-7816.
Miller, J.R. and Beitz, J.V. 1981. *J. Chem. Phys.*, 74, 6746.
Miller, J.R., Beitz, J.V. and Huddleston, R.K. 1984. *J. Am. Chem. Soc.*, 106, 5057-5068.
Oevering, H., Paddon-Row, M.N., Heppener, M., Oliver, A.M., Cotsaris, E., Verhoeven, J.W. and Hush, N.S. 1987. *J. Am. Chem. Soc.*, 109, 3258-3269.
Pasman, P., Mes, G.F., Koper, N.W. and Verhoeven, J.W. 1985. *J. Am. Chem. Soc.*, 107, 5839-5843.
Warman, J.M., de Haas, M.P., Paddon-Row, M.N., Cotsaris, E., Hush, N.S., Oevering, H. and Verhoeven, J.W. 1986. *Nature*, 320, 615-616.

ELECTRON TRANSFER PHOTOSENSITIZED BY ZINC PORPHYRINS IN REVERSED MICELLES

Sílvia M.B. Costa

Centro de Química Estrutural
Complexo I, Instituto Superior Técnico
Av. Rovisco Pais
1096 Lisboa Codex, Portugal

ABSTRACT

Kinetic studies of photosensitized electron transfer by a non water-soluble zinc tetraphenylporphyrin (ZnTPP) and water-soluble, zinc tetramethylpyridilporphyrin (ZnTMePyP^{4+}), zinc tetraphenylporphyrin sulphonate (ZnTPPS^{4-}) and the sodium salt of zinc tetraphenylporphyrin 4-carboxylic acid (ZnTPPCOONa)$_4$ to duroquinone (DQ), methyl viologen (MV^{2+}) and anthraquinone-2-sulphonate (AQS$^-$) in reversed micelles of benzyldimethyl-n-hexadecylammonium chloride were carried out using fluorescence quenching and laser flash photolysis techniques varying the water content as expressed by the relation $w_o = [H_2O]/[S]$. Neutral quenchers, such as (DQ), quench dynamically only porphyrin molecules located in the organic phase or at the interface. A quencher bound at the positive interface, such as (AQS$^-$) interacts statically with porphyrin molecules at the interface but does not quench porphyrin molecules in the organic phase and the quenching data can be interpreted in terms of an active-sphere quenching model. In contrast to this, if porphyrin molecules are in the water pool, the quenching is dynamic and its efficiency increases with the size of the water pool. The quencher located in the water pool (MV^{2+}), not bound at the interface interacts only with porphyrin molecules also in the water pool. At high water content a statistical homogeneous distribution is observed but the quenching efficiency is much greater than in the aqueous solution, whereas at lower water content the water pool is rigid, the interaction is static and a Poisson distribution of the quencher is observed. Laser flash photolysis was carried out at $w_o = 22$ and again statistical kinetics could be applied when (MV^{+2}) was the quencher, whereas a Poisson distribution was used for (AQS$^-$) quencher.

INTRODUCTION

Micelles and microemulsions have been explored as membrane mimetic systems since they possess charged microscopic interfaces which act as barriers to the charge recombination process (Fendler et al., 1980; Hurst et al., 1983). Namely, the influence of the location of the sensitizer on photoinduced electron transfer kinetics and on charge separation between photolytic products in reversed micelles has been studied (Pileni et al., 1985).

Micellar media, relatively to the homogeneous one, change the distribution and effective concentration of species involved in a given reaction. Kinetic studies are therefore simplified by carefully controlling the experimental conditions.

Absorption and emission spectra of zinc tetraphenylporphyrin derivatives soluble and insoluble in water showed an equilibrium between two spectroscopically different porphyrin species incorporated in reversed micelles of benzyldimethyl-n-hexadecylammonium chloride in benzene formed by varying the water content given by $w_o = [H_2O]/[Surf]$. (Costa et al., 1985 and 1986).

Time-resolved fluorescence emission techniques provided additional information on the location of those porphyrins in these reversed micelles (Costa et al., 1986).

The purpose of this paper is to show that the selective compartmentalization achieved in reversed micelles affects drastically the spectroscopic and redox behaviour of solubilized probes. This effect can be used to monitor electron transfer kinetics.

MATERIALS AND METHODS

ZnTPP (Hambright, Washington D.C.), duroquinone (DQ) (Aldrich) and methyl viologen (MV^{+2}) (BDH CHEMICALS) were used without further purification. The sodium salt of anthraquinone 2-sulphonate (AQS) (BDH CHEMICALS) was recrystalized from methanol. Benzyldimethyl-n-hexadecylammonium chloride (BHDC)(BDH CHEMICALS) was dried over P_2O_5 and stored under vacuum. The solvent, benzene (pro-analysis)(BDH CHEMICALS) was dried and distilled over sodium wire. Stock solutions of reversed micelles were prepared by adding dropwise the calculated amount of water to the appropriate solution of BHDC. The surfactant concentration used throughout the experiments was 0.1 mol dm^{-3}.

Zinc tetramethylpyridilporphyrin tetraiodide was prepared by refluxing in water the free base meso-tetra(4-N-methylpyridine)porphyrin tetraiodide (Strem Chemicals) with finely divided zinc oxide (Herrmann et al., 1978). An identical procedure was used to obtain zinc tetrasulphonate phenylporphyrin from the free base tetrasodium-meso-tetra(4-sulphonatephenyl)porphyrin (12 hydrate)(Strem Chemicals). The purity of these porphyrins was checked by comparing the absorption spectra of the metallated porphyrin with those of the free base.

Absorption spectra were recorded with a Perkin-Elmer Lambda 5 spectrophotometer in 1 cm quartz cells. The emission spectra were recorded with an MPF-3 Hitachi spectrofluorimeter, also in 1 cm quartz cells.

Laser flash photolysis was performed with a Nd-YAG Laser from Quantel which was frequency doubled with a KDP-Crystal. Samples were excited at 532 nm and the transients formed monitored with a pulsed 250 W Xenon lamp and recorded using a monochromator and an R128 photomultiplier connected with a transient digitizer linked to a PDP11-73 minicomputer.

RESULTS AND DISCUSSION

Absorption and Emission

The absorption and emission spectra of ZnTPP and derivatives in reversed micelles of BHDC depend strongly on the molar ratio (w_o) of water and surfactant (Costa et al., 1985). Indeed, for non water-soluble porphyrins, drastic changes are observed in the spectra upon addition of water and in the limit the spectra are identical with those obtained in a homogeneous solution of benzene. Likewise, for water-soluble zinc porphyrins, at low water content, these molecules are associated with the interface, but spectral shifts are also observed and at high water content the spectra tends to the one in water. However, charged surfactant head groups at the interface interact electrostatically with ionic porphyrins and determine their location in the water pool. Therefore, negatively charged porphyrins remain near the interface even at high water content, whereas positively charged porphyrins are pushed away from the interface (Costa et al., 1986).

Spectral changes with the water content are attributed to the existence of aggregates which induce different environments surrounding the porphyrin and not to dimer formation (Brochette et al., 1985). The observation of an isosbestic point implies the existence of two different species of porphyrin, one associated with the interface and the other corresponding to the free porphyrin in the bulk organic solvent or water. This equilibrium is expressed by

$$P_i + yH_2O_{add} \rightleftarrows P_e + yH_2O_{pool}$$

where P_i and P_e stand for associated porphyrin at the interface and free porphyrin in the bulk solvent, respectively.

The fraction of aggregates, p, is expressed by eq (1)

$$p = \frac{[P_i]}{[P_t]} = \frac{1}{K[H_2O]^y + 1} \qquad (1)$$

where the equilibrium constant is given by $K = [P_e]/[P_i][H_2O]^y$ and y is the number of molecules of H_2O which intervene in the equilibrium. Since for a given w_o the two spectroscopically distinct porphyrins are always present (Costa et al., 1986), it is necessary to know accurately the distribution function $p(w_o)$ to evaluate the extent of fluorescence quenching in the presence of quenchers and when determining kinetic parameters.

The emission spectra studied as a function of w_o enable the determination of the equilibrium constant K referred to, using eq (2)

$$(I_o/I_f) - 1 = I_o/(I_e + p(w_o) K (H_2O)^y) \qquad (2)$$

where the distribution function $p(w_o) = 1/(1 + K (H_2O)^y)$, I_o and I_e represent the fluorescence emission intensity when all the porphyrin molecules are at the interface or in the water pool (or organic phase) respectively, and I_f is the fluorescence emission intensity at a given w_o and wavelength. The data extracted by studying the effect of water content w_o on the fluorescence emission of several porphyrins are presented on Table I

TABLE 1 Equilibrium constants K and number of molecules of water y for several zinc porphyrins in reversed micelles of BHDC.

PORPHYRIN	K (dm^3 mol^{-1})	y
ZnTPP	30.0 ± 2	2.8 ± 0.1
ZnTPP($C_{12}H_{25}$)	60.0 ± 8	2.4 ± 0.12
ZnTPPR$_4$ (n = 1)	14.0 ± 1.2	2.3 ± 0.08
ZnTPPR$_4$ (n = 5)	28.3 ± 0.2	2.7 ± 0.2
ZnTPPR$_4$ (n = 6)	44.0 ± 8	3.2 ± 0.2
ZnTPPR$_4$ (n = 17)	36.0 ± 9	2.9 ± 0.2
ZnTPP(COOMe)$_4$	6.5 ± 1.3	2.4 ± 0.2
ZnTPPS^{4-}	5.4 ± 0.7	2.4 ± 0.2
ZnTMePyP^{4+}	6.6 ± 0.9	2.3 ± 0.2
ZnTPP(COONa)$_4$	0.25 ± 0.3	6.7 ± 0.7
ZnTPP(COONa)$_3$ - C_3 - ZnTPP	2.5 ± 0.7	2.5 ± 0.6

$R \equiv O(CH_2)_n CH_3$

Examination of Table I shows that the water-soluble porphyrins stay nearer to the interface, this being favoured when they bear a negative charge which is attracted to the positively charged interface. On the other hand, the non water-soluble porphyrins, due to a greater hydrophobicity, are displaced from the interface more easily, which is reflected in larger values of the equilibrium constants.

Fluorescence quenching kinetics

Fluorescence quenching studies are an important tool for the elucidation of the structure and dynamics of these aggregates. The kinetics observed are strongly dependent on the specific probe and quencher location.

Probe at the interface

At very low water content (w_o = 0-2) the porphyrins (ZnTPP and ZnTPPS^{4-}) are located at the interface and are dynamically quenched by duroquinone (DQ), which is solubilized in the organic phase, with observed rate constants of k_q = 9.1 x 10^9 dm^3 mol^{-1} s^{-1}, ZnTPP/DQ, w_o = 0 and k_q = 1.9 x 10^9 dm^3 mol^{-1} s^{-1}, ZnTPPS^{4-}/DQ, w_o = 2. However, a quencher bound at the interface like anthraquinone-2-sulphonate (AQS$^-$) interacts statically with ZnTPP at the interface as shown by steady-state and transient fluorescence techniques (Table 2)

TABLE 2 Analysis of ZnTPP decay in BHDC benzene (w_o = 0) in the presence of AQS$^-$

[AQS$^-$]/10^{-3} mol dm^{-3}	τ_1/ns	τ_2/ns
0.0	1.08	–
1.0	1.10	0.33
2.0	1.10	0.11
3.0	1.06	0.11
3.5	1.05	0.12
4.0	0.99	0.11

The fluorescence decay has a major component of ≈ 1 ns which is likely to correspond to unquenched porphyrin, whereas a much shorter component of ~ 100 ps can also be detected which is attributed to quenched porphyrin by quinone molecules which "sit" within a distance corresponding to the radius of an active-sphere of quenching (Perrin, 1924).

Probe in the water pool

We have studied the fluorescence quenching of ZnTMePyP^{4+} which is located in the water pool, using different quenchers such as AQS$^-$ and MV^{2+}.

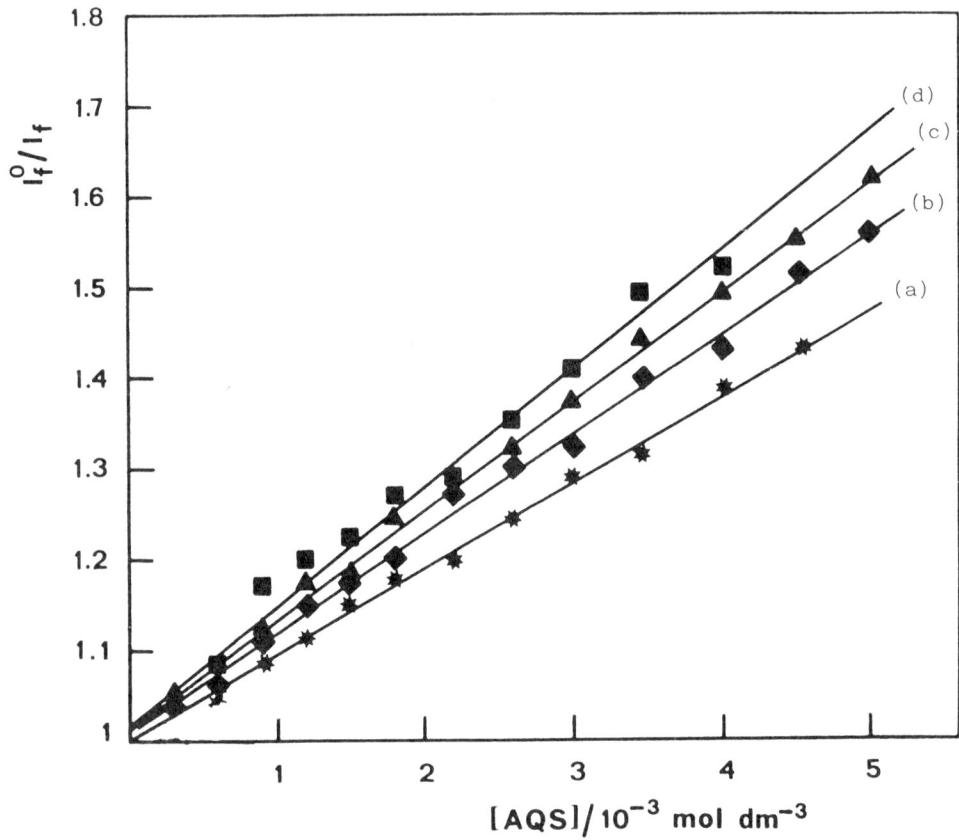

Fig. 1 Fluorescence quenching of ZnTMePyP^{4+} by AQS$^-$ in reversed micelles of BHDC ([S] = 0.1 mol dm^{-3}) at different w_o values: a) 6, b) 11, c) 22, d) 28.

It can be seen that when the quencher is bound at the interface (AQS$^-$) only quencher molecules which can come into the water pool are efficient in the quenching process which is therefore dependent on the size of the water pool, related to the value of w_o. On the contrary, if quencher molecules are solubilized in the water pool (MV^{+2}), the quenching efficiency is the same above $w_o > 2$ and much greater than in pure water.

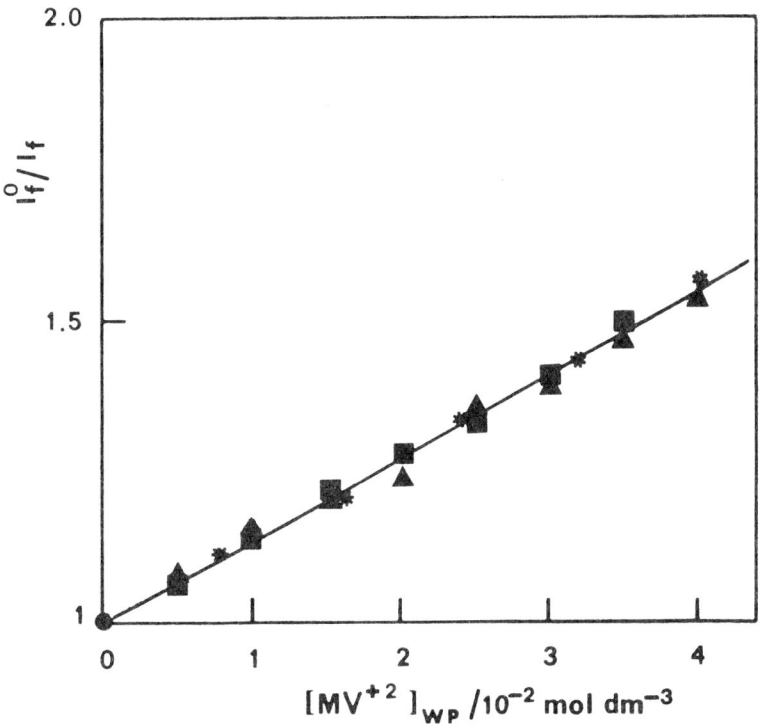

Fig. 2 Fluorescence quenching of ZnTMePyP$^{4+}_{-3}$ by MV$^{+2}_{wp}$ in reversed micelles of BHDC ($|S| = 0.1$ mol dm^{-3}) at several w_o values: 4, 12, 22.

The quenching of ZnTPP(COONa)$_4$ by MV^{+2} in these micelles is also dynamic, unlike that found in water for this probe/quencher pair where complete aggregation is observed. Nevertheless, the quenching rate constants are also somewhat dependent on the value of w_o, as can be seen on Table 3.

TABLE 3 Variation of quenching rate constants for the system ZnTPP(COONa)$_4$/MV^{+2} in BHDC with w_o.

$k_q/10^9$ dm^3 mol^{-1} s^{-1}	8	10	12	14	16	18	22
w_o	1.22	1.24	1.57	1.63	1.57	1.67	1.62

Laser Flash Photolysis

In order to obtain the radical anion of the acceptor, experiments using the technique of Laser Flash photolysis were carried out in the systems ZnTPP/AQS$^-$ or MV^{+2} (w_o = 0; 10) and ZnTMePyP^{4+}/AQS$^-$ or MV^{+2} (w_o = 22). Comparison of Fig. 3 and Fig. 4 in the absence and presence of AQS$^-$ clearly shows the appearance of a species at 500 nm which is attributed to the radical anion. The rate of recombination is delayed at lower water content.

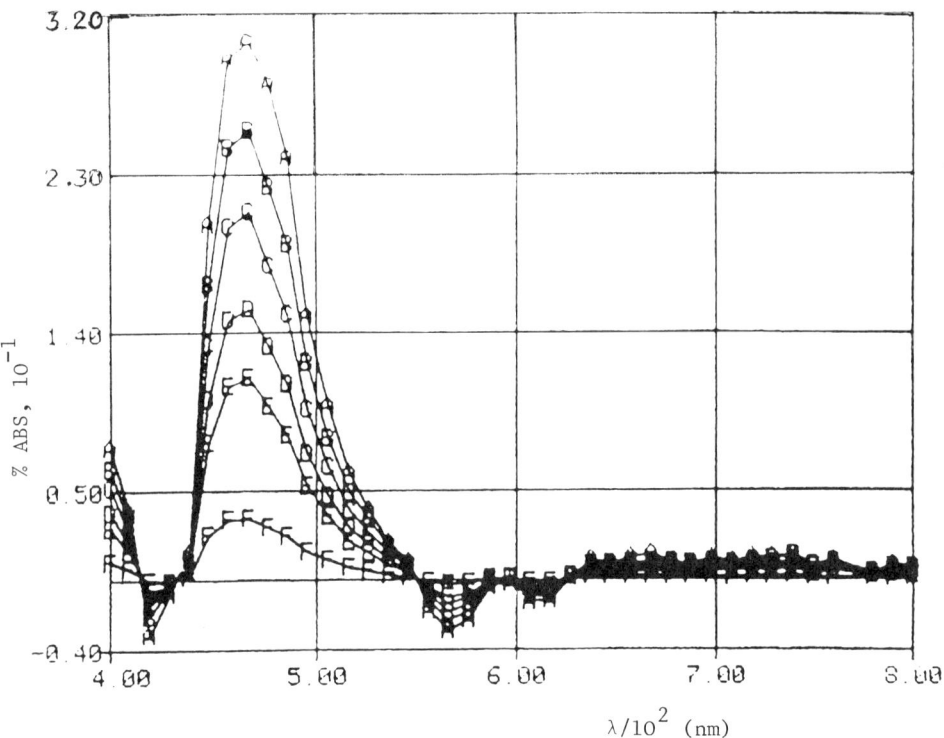

Fig. 3 Time resolved spectra of the triplet of ZnTPP in reversed micelles of BHDC (w_o = 0) in the absence of AQS$^-$.
Time (μsec): A 2.00; B 14.00; C 26.00; D 44.00; E 64.00; F 169.00

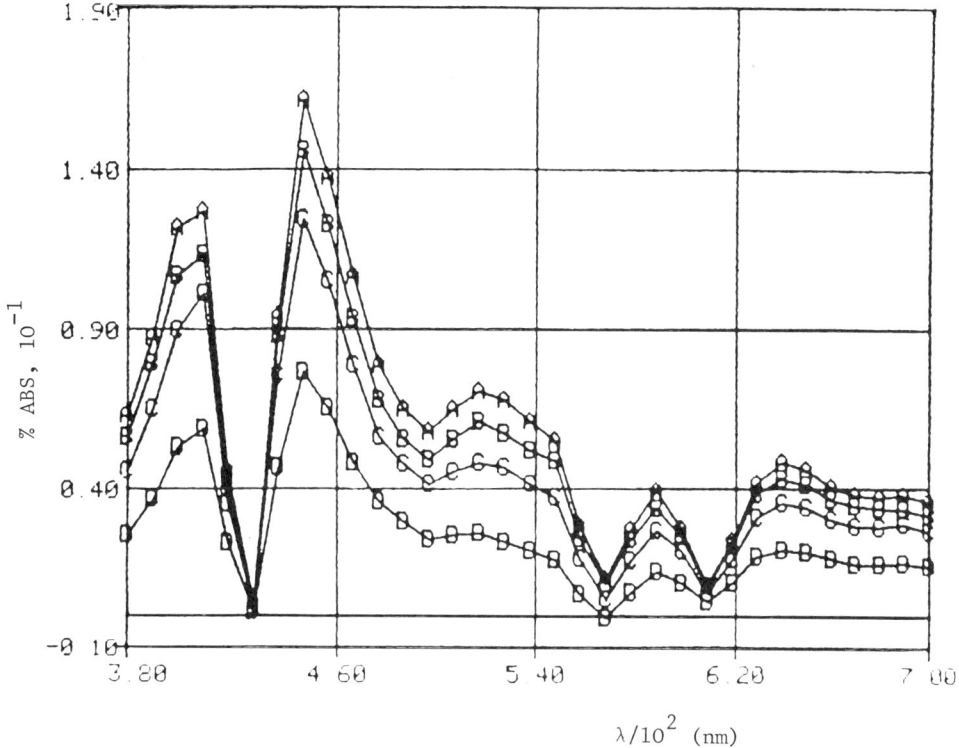

Fig. 4 Time resolved spectra of the triplet of ZnTPP in reversed micelles of BHDC (w_o = 0) in the presence of AQS$^-$.
Time (µsec): A 1.00; B 13.00; C 37.00; D 134.50

Methyl viologen MV^{+2} does not quench the triplet of ZnTPP in these reversed micelles, but quenches the triplet of ZnTMePyP^{4+} with a rate constant of k_q = (2 ± 0.3) x 10^8 dm^3 mol^{-1} s^{-1}, but it was not possible to detect the radical anion.

CONCLUSION

This study showed that the properties of reversed micelles depend critically on the water content below the maximum hydration number. Furthermore, they offer a means of compartmentalizing the reaction partners and thus impose a barrier on the rate of recombination of transient species formed by direct electron transfer.

ACKNOWLEDGEMENTS

This work was supported by Instituto Nacional de Investigação Científica (I.N.I.C.).

We are indebted to Dr. A. Harriman and Dr. M. Nango for the generous gifts of $ZnTPP(C_{12}H_{25})$ and $ZnTPPR_4$ (n = 1, 5, 6, 17) and $ZnTPP(COOMe)_4$, $ZnTPP(COONa)_4$ and $ZnTPP(COONa)_3 - C_3 - ZnTPP$.

REFERENCES

Brochette, P. and Pileni, M.P. 1985. Nouv. J. Chim., 9, 557.
Costa, S.M.B., Aires-Barros, M.R. and Conde, J.P. 1985. J. Photochem., 28, 153.
Costa, S.M.B. and Brookfield, R.L. 1986. J. Chem. Soc. Faraday Trans. 2, 82, 991.
Costa, S.M.B., Lopes, J.M.F.M. and Martins, M.J.T. 1986. J. Chem. Soc. Faraday Trans. 2, 82, 2371.
Herrmann, O., Husain Mehdi, S. and Corsini, A. 1978. Can. J. Chem., 56, 1084.
Hurst, J.K., Lee, L.Y.C. and Grätzel, M. 1980. J. Am. Chem. Soc., 105, 7048.
Infelta, P.P., Grätzel, M. and Fendler, J.H. 1980. J. Am. Chem. Soc., 102, 1479.
Perrin, F., 1924, C.R. Sceances Acad. Sci., 178, 1978.
Pileni, M.P., 1981. Chem. Phys. Lett., 81, 603.

EFFICIENT VISIBLE LIGHT SENSITIZATION OF TiO$_2$ BY SURFACE COMPLEXATION WITH TRANSITION METAL CYANIDES

E. Vrachnou[*], N. Vlachopoulos and M. Grätzel

Institut de chimie physique, Ecole Polytechnique Fédérale de Lausanne, CH-1015 Lausanne, Switzerland.
[*]On leave from N.R.C. "Demokritos", Aghia Paraskevi, 153 10 Athens, Greece.

ABSTRACT

Charge-transfer complexes absorbing strongly in the visible spectrum are formed on the surface of TiO$_2$ by adsorption of the transition metal cyanides Fe^{II}(CN)$_6^{4-}$, Ru^{II}(CN)$_6^{4-}$, Os^{II}(CN)$_6^{4-}$, Re^{II}(CN)$_6^{4-}$, Mo^{IV}(CN)$_8^{4-}$ and W^{IV}(CN)$_8^{4-}$. This results in visible light photosensitization of TiO$_2$, due to electron injection from the excited state of the complexes to the conduction band of the semiconductor. Photoelectrochemical systems with photoelectrodes of polycrystalline TiO$_2$ derivatized with the above complexes give quantum yields of up to 37% upon illumination at the absorption peaks of the complexes, around 420 nm. The photoresponse is extended up to 700 nm.

INTRODUCTION

Semiconductors have proven to be very important factors in the photovoltaic conversion of solar radiation into electric energy. They can also be used as photoelectrodes in photoelectrochemical cells producing power or fuels, and in photocatalytic reactions of high specificity, though these applications are still at the experimental stage.

For photoelectrochemical functions the semiconductors should have the following properties, in addition to their semiconducting properties:
(1) the energy of the band gap should correspond to the energy of the visible part of the solar radiation, which represents a significant portion of the solar spectrum, and (2) they should be resistant against corrosion or dissolution caused by electrolyte solutions in processes under illumination as well as in darkness.

Up to now, all known semiconductors with low band gap are not resistant enough in solution and they tend to dissociate, especially when used as photoanodes. Stable semiconductors have a wide band gap and therefore can only be active in UV light. TiO$_2$ falls in this category, with a band gap of 3 eV. Upon illumination with UV light, electrons are excited to the conduction band, leaving holes in the valence band. These electrons and holes can act as strong oxidants and reductants respectively, initiating redox reactions which can lead to various important processes, such as water cleavage, reduction of carbon dioxide, etc. (Grätzel M., 1983). In visible light TiO_2 is inert. The spectral response of TiO_2 can be expanded, though, to the visible region by use of photosensitizers, i.e. strongly colored substances absorbing highly in the visible region, which upon excitation can inject an electron into the conduction band of TiO_2. The photosensitization of TiO_2 has been the object of intensive investigation in recent years. (Gerischer H., 1972). Figure 1 illustrates the effect of sensitization of the semiconductor photoanode in a photoelectrochemical cell.

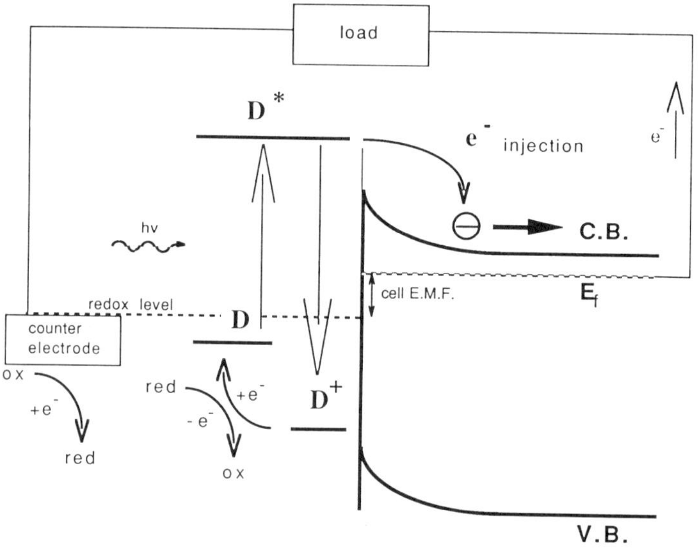

Fig. 1 Regenerative dye-sensitized photoelectrochemical cell.

Here we report the efficient spectral sensitization of TiO_2 by surface derivatization with a series of transition metal cyanides. The sensitization of TiO_2 when treated with a ferrocyanide solution is described in a recent publication from this laboratory (Vrachnou E., Vlachopoulos N., and Grätzel M., 1987). There is evidence that the active species formed on the surface is a titanium analogue of Prussian blue.

Prussian blue, $Fe^{III}_4[Fe^{II}(CN)_6]_3 \cdot xH_2O$, was first prepared in 1710 by Diesbach in Berlin, but its structure was first elucidated by Keggin and Miles in 1936, refined by Ludi and Güdel in 1973, and is still subject to investigation in its details (Buser et.al. 1977; Rasmussen and Meyers, 1984).

Prussian blue is the prototype of a series of complexes of the general formula $A_m[B(CN)_1]_n \cdot x\ H_2O$ or $K_nA[B(CN)_1] \cdot x\ H_2O$, where A and B are transition metal ions, and K is an alkali ion. A and B can be the same metal ion in different oxidation states, as in Prussian blue. A is coordinated through nitrogen to the cyano group, while B is coordinated through carbon. (Ludi and Güdel, 1973; Rasmussen and Meyers, 1984).

Many of these complexes are deeply colored and their electronic spectra contain very strong intervalence transfer bands in the visible, due to electron transfer from B to A (Robin and Day, 1967). They also show magnetic coupling and they have semiconducting properties. They are stable in air and in acid solutions, but they dissociate in alkali. They can be reversibly oxidized, whereby B gives up an electron and goes to a higher oxidation state, and the intense color disappears.

Because of their properties they are widely used in electrochemistry, in electrophotography, as well as in photocatalysis and photochemical water cleavage. Thus Hennig and Rehorek in 1986 reported the sensitization of photochemical reactions by Prussian blue analogues of molybdenum octacyanide, and Kaneko and his group in 1984 found that water can be photolyzed with visible light in presence of Prussian blue and tris (2,2'-bipyridine) ruthenium (II) complex.

REACTIONS OF TiO_2 SURFACES WITH FERROCYANIDE IONS

An orange color is formed upon mixing deaerated TiO_2 colloid solutions with $K_4Fe^{II}(CN)_6$ solutions at pH $\leq$ 5. The absorption spectrum of the mixture has a broad absorption band with a peak at 420 nm (Figure 2). The color is more intense between pH 2 and 3. The presence of oxygen inhibits the reaction. The oxidized complex, $K_3Fe^{III}(CN)_6$, is inactive. The same color is formed on the surface of TiO_2 powders (Degussa P25 or anatase) under similar conditions.

Orange-brown precipitates are formed upon mixing acid aqueous TiO^{2+} solutions with solutions of $K_4Fe^{II}(CN)_6$ or $TiCl_3$ solutions with solutions of $K_3Fe^{III}(CN)_6$. The products of these reactions have already been studied by X-rays and Mössbauer spectroscopy (Maer et.al. 1968) and were found analogous to Prussian blue, with the formula $Ti^{IV}[Fe^{II}(CN)_6]$.

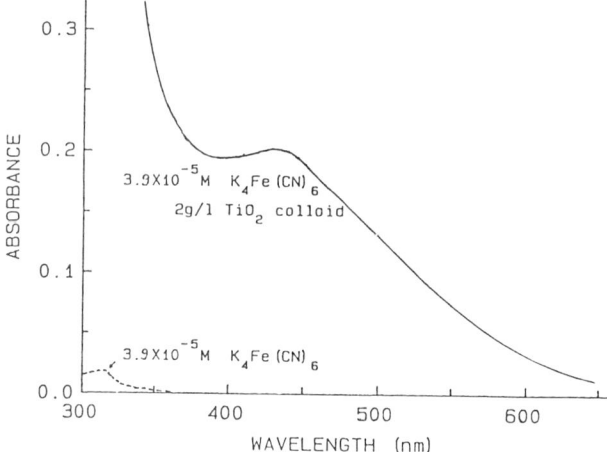

Fig. 2 Absorption spectra. (---)$K_4Fe(CN)_6$, 3.9×10^{-5} M, absorbs weakly in the ultraviolet. (——) complex formed on TiO_2 colloids; the differential spectrum is taken relative to a TiO_2 reference, and shows the strong absorption by the complex in the visible spectrum.

From the similarity of the electronic absorption and reflectance spectra (Figure 3) and also of the position of the C≡N stretching vibrations in the IR spectra of TiO_2 powders treated with $Fe(CN)_6^{4-}$ (Table 1) we can conclude that the complexes formed on the TiO_2 surfaces are of the same type.

TABLE 1 C-N stretching vibrations of the cyano-group

Complex	$\nu_{C≡N}(cm^{-1})$
$K_4[Fe^{II}(CN)_6]$	2040
$K_3[Fe^{III}(CN)_6]$	2114
$Fe_4^{III}[Fe^{II}(CN)_6]_3$	2080
$Ti^{IV}[Fe^{II}(CN)_6]$	2078
$TiO_2Fe(CN)_6^{4-}$ [a]	2078

[a] Determined from FTIR spectra taken by Desilvestro (unpublished results).

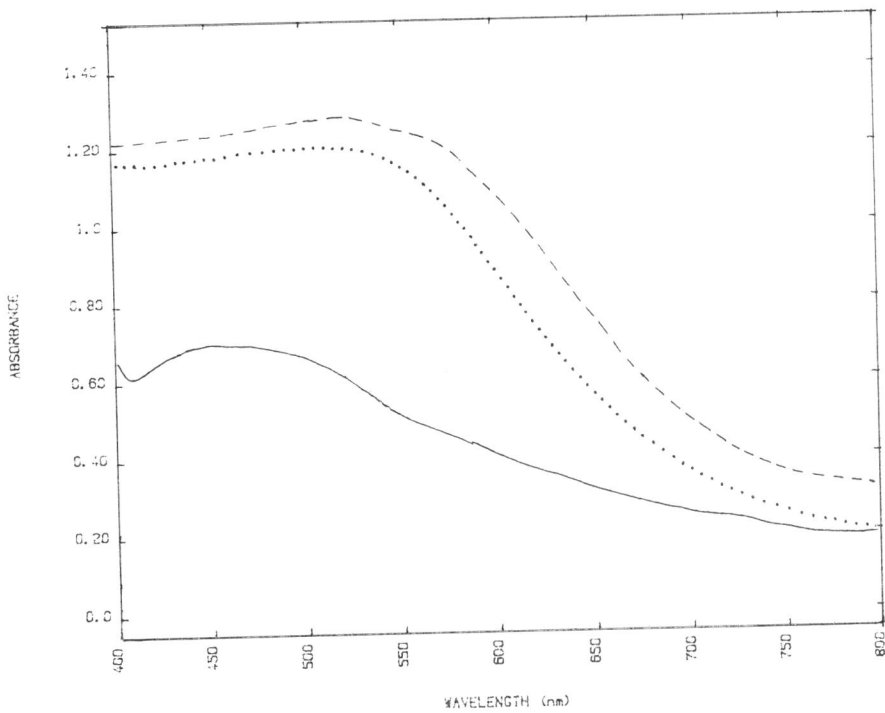

Fig. 3 Reflectance spectra of TiO_2 powder treated with $Fe^{II}(CN)_6^{4-}$ (———), and of the complexes formed from Ti^{3+} and $Fe^{III}(CN)_6^{3-}$ (- - -), and from TiO^{2+} and $Fe^{II}(CN)_6^{4-}$ (...).

The extinction coefficient of the complex formed in TiO_2 colloid reaction mixtures at 430 nm has the value of $5.2 \times 10^3 \, dm^3 \, mol^{-1} \, cm^{-1}$.

PHOTOCHEMICAL BEHAVIOR OF TiO_2 SURFACES DERIVATIZED WITH FERROCYANIDE

The visible light sensitization of ferrocyanide derivatized TiO_2 colloid was investigated by laser flash photolysis (Figure 4). On illumination by a light pulse of 20 nsec. duration, with a wavelength of 530 nm, longer than that for the absorption edge of TiO_2 but within the absorption band of the complex, a bleaching of the chromophore takes place. The absorption at 480 nm is significantly reduced within the 20 nsec. duration of the laser pluse and recovers subsequently with a half-lifetime of 3 μs. This behavior shows that charge transfer with electron injection into TiO_2 conduction band levels does in fact occur (Reaction 1). The rate constant for the back electron transfer is ca. $2 \times 10^5 \, s^{-1}$, which is typical for

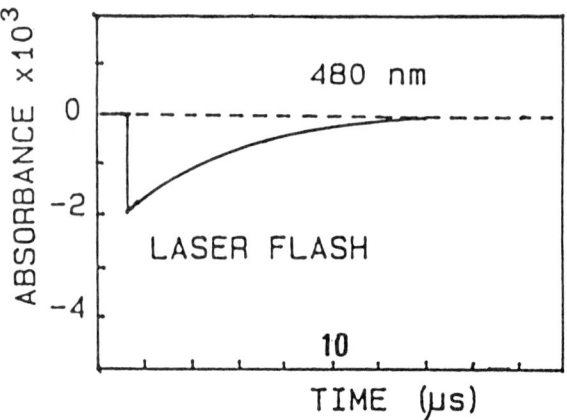

Fig. 4 530 nm laser flash photolysis on colloidal $TiO_2/Fe(CN)_6^{4-}$ complex. Absorption is measured at 480 nm, and indicates the lifetime of the photoexcitation.

the recapture of conduction band electrons by oxidized sensitizers adsorbed at the surface of colloidal TiO_2 particles (Reaction 2)(Moser and Grätzel, 1984 ; Desilvestro et.al., 1985).

$$[Fe(CN)_6^{4-}]_{TiO_2} \xrightarrow{h\nu} e_{cb}^-(TiO_2) + [Fe(CN)_6^{3-}]_{TiO_2} \quad (1)$$

$$[Fe(CN)_6^{3-}]_{TiO_2} + e_{cb}^-(TiO_2) \longrightarrow [Fe(CN)_6^{4-}]_{TiO_2} \quad (2)$$

Photosensitization of TiO_2 powders (Degussa P25 with 0.5% Pt on the surface) treated with $Fe(CN)_6^{4-}$, was observed in preliminary experiments where lactic acid was oxidized to pyruvic acid with simultaneous hydrogen evolution, upon visible light irradiation of suspensions of the derivatized powder in dilute lactic acid solutions.

DERIVATIZATION OF TiO_2 SURFACES WITH METAL CYANIDES

Acid solutions of $Ru^{II}(CN)_6^{4-}$, $Os^{II}(CN)_6^{4-}$, $Re^{II}(CN)_6^{4-}$, $Mo^{IV}(CN)_8^{4-}$, and $W^{IV}(CN)_8^{4-}$ have a similar effect on TiO_2 as $Fe^{II}(CN)_6^{4-}$. They also form orange charge transfer complexes under the same conditions. The absorption spectra of the complexes with $Re^{II}(CN)_6^{4-}$, $W^{IV}(CN)_6^{4-}$ resemble the spectrum of $TiO_2-Fe^{II}(CN)_6^{4-}$, while those with $Os^{II}(CN)_6^{4-}$ and $Ru^{II}(CN)_6^{4-}$ are shifted towards the UV.

TiO$_2$ powders (Degussa P25 or anatase) also become intensely colored when suspended in solutions of the cyanides. Reflectance spectra are shown in Figure 5.

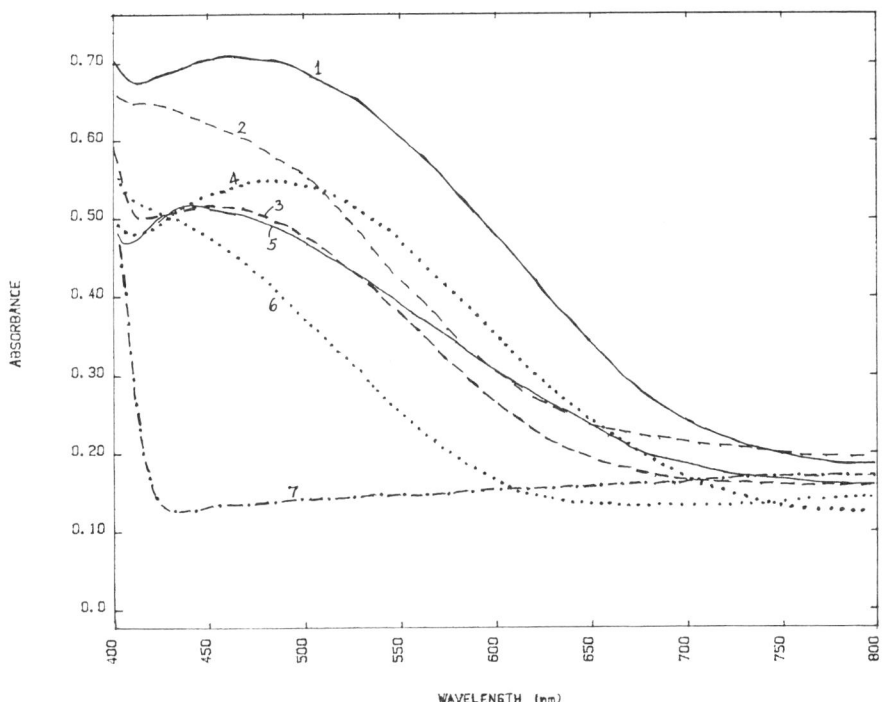

Fig. 5 Reflectance spectra of TiO$_2$ powder treated with Fe^{II}(CN)$_6^{4-}$, (1), Os^{II}(CN)$_6^{4-}$ (2), Mo^{IV}(CN)$_8^{4-}$ (3), W^{IV}(CN)$_8^{4-}$ (4), Re^{II}(CN)$_6^{4-}$ (5), Ru^{II}(CN)$_6^{4-}$ (6), and untreated TiO$_2$ (7).

Photoelectrochemical Experiments

For the photoelectrochemical experiments a three-compartment cell was used, equipped with a quartz window. The working electrodes were cylindrical titanium rods (geometric area 0.28 cm^2) onto which polycrystalline anatase layer (thickness ca. 20 μm, surface roughness factor ca. 280) was deposited, according to the method of Stalder and Augustynski (1979), modified in this laboratory (Vlachopoulos et.al., 1987). The electrodes were immersed for a few hours in 10^{-2}M solutions of the corresponding cyanides in HClO$_4$ 10^{-2} or 10^{-3}M, then washed and dried. The same charge transfer complexes were formed on the electrode surface. Perchloric acid solutions (10^{-2} to 10^{-3}M) were used for the experiments and hydroquinone was added as supersensitizer.

For all the derivatized electrodes the photocurrent onset is at a wavelength greater than 600 nm, where the photon energy is 2.0 eV in contrast to the 3.1 eV band gap of the semiconductor. For untreated TiO_2 electrodes the photocurrent action spectrum follows the absorption edge of the semiconductor, with little response in the visible. The onset potential in all cases is the same as with the untreated electrode (Figure 6).

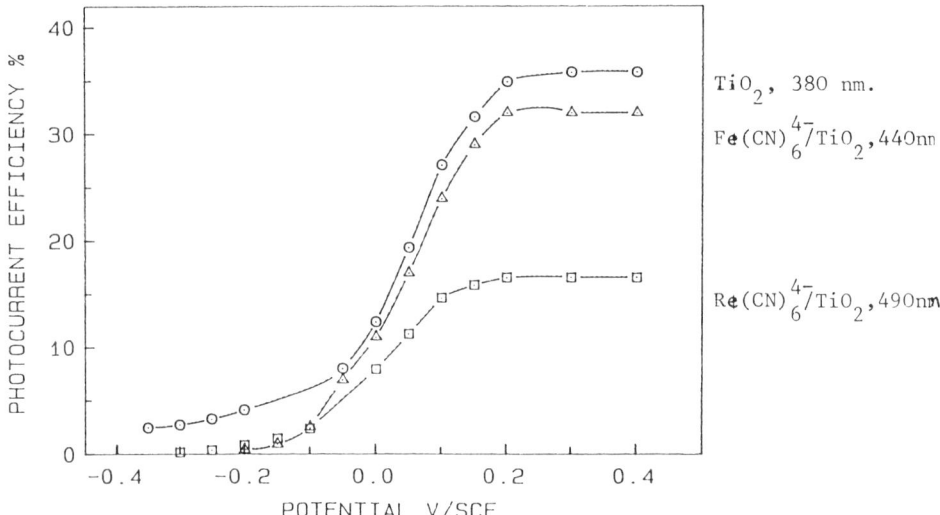

Fig. 6 Photocurrent/potential characteristics for TiO_2 (○), and rhenium (■) and iron (▲) cyanide complexed TiO_2 electrodes. The TiO_2 was illuminated by UV (380 nm), the others by visible light.

All the response spectra were taken in standardized solutions, with the electrode potential under potentiostatic control at +0.2V vs SCE.

As can be seen from the action spectra in Figures 7 - 9, of all the transition metal cyanides tested to date, the ferrocyanide derivatization of TiO_2 gives the most effective sensitization: at the peak of the photocurrent action spectrum at a wavelength of 420 nm, corresponding to the absorption maximum of this complex, the quantum efficiency for electron transfer is 37% (monochromatic incident light intensity 0.84 $W.m^{-2}$, photocurrent 10.4 $\mu A.cm^{-2}$).

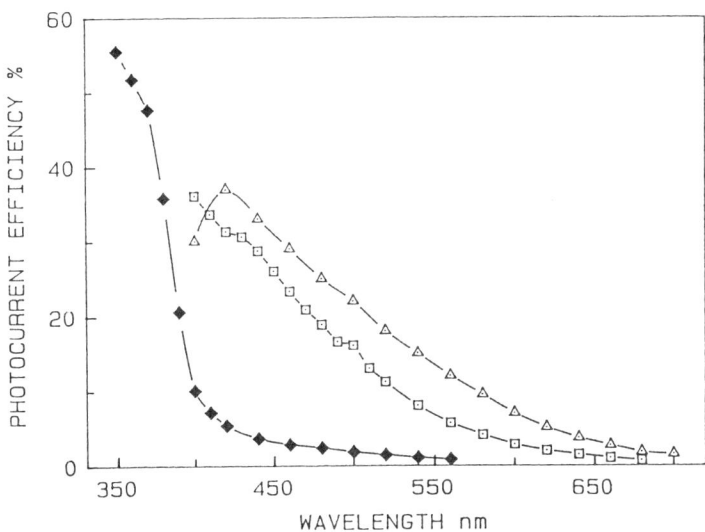

Fig. 7   Photocurrent efficiency (charge transfer/photon flux) of rhenium (□) and iron (△) cyanide complexed TiO_2 photoelectrodes with that of untreated TiO_2 (◆) for comparison. The extension of the spectral response into the visible is evident.

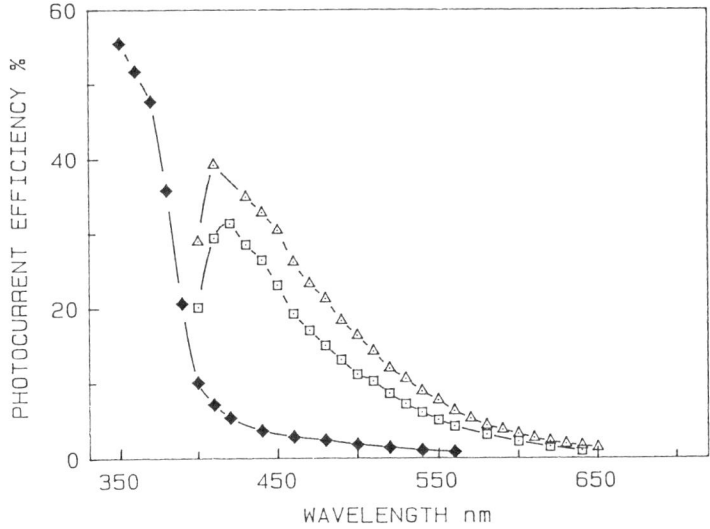

Fig. 8   Photocurrent quantum efficiency of ruthenium (△) and osmium (□) cyanide complexed TiO_2 electrodes compared with that of untreated TiO_2(◆).

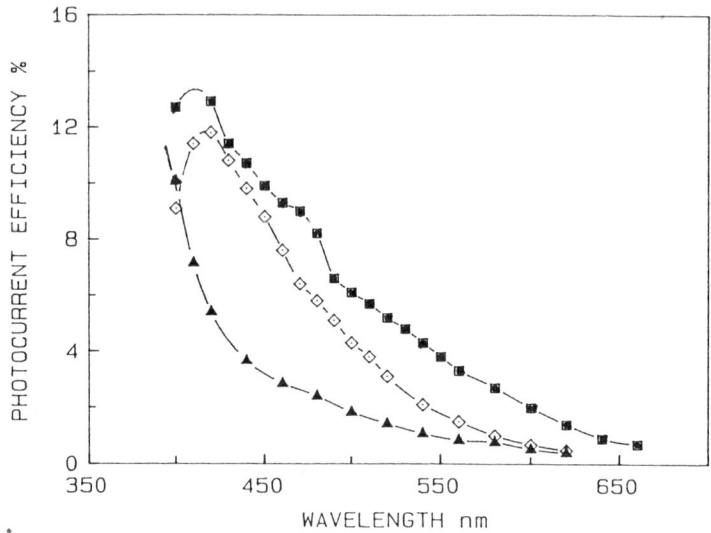

Fig. 9 Photocurrent quantum efficiency of tungsten (■) and molybdenum (◇) cyanide-complexed TiO_2 electrodes compared with that of untreated TiO_2 (▲).

CONCLUSION

Efficient photosensitization of TiO_2 surfaces by metal cyanide complexes occurs through formation of titanium analogues of Prussian blue compleyes, which absorb strongly in the visible. The photosensitization has been established for TiO_2 in colloid and powder form, as well as for extended surface area ("fractal") electrodes. Optimization of derivatization procedures and selection of the most effective electrode, using the ferrocyanide derivatization, has given a quantum efficiency of 37% at 420 nm, while the photoresponse extends to 700 nm. As part of the ongoing investigation, regenerative cells will be developed, in which the energy is derived as electricity.

ACKNOWLEDGMENT

This work was supported by the Gas Research Institute, Chicago, Ill., USA (subcontract with the Solar Energy Research Institute, Golden, Colorado) and the Nationaler Energie Forschungs Fonds, Switzerland.

The authors are grateful to Professor Martin Fleischmann for inspiring discussions concerning the fractal character of the TiO_2 layer, to Dr P.P. Infelta for his valuable assistance with the laser experiments, to Mr P. Liska for the preparation of the TiO_2 electrodes, and to Ms A. Farina for the photocatalysis experiments.

REFERENCES

Buser, E.J., Schwarzenbach, D., Petter W. and Ludi A. 1977. The crystal structure of Prussian blue. Inorg. Chem., 16, 2704-2710.
Desilvestro, J., Grätzel M., Kavan, L., Moser, J. and Augustynski, J. 1985. Highly efficient sensitization of titanium dioxide. J.Am.Chem.Soc., 107, 2988-2990.
Diesbach, 1710. Miscellanea Berolinensia ad incrementum scientiarum, Berlin, pp 377-379.
Gerischer, H. 1972. Electrochemical techniques for the study of photosensitization. Photochem. Photobiol., 16, 243-260.
Grätzel, M. (Editor). 1983. "Energy Resources through Photochemistry and Catalysis" (Academic Press, New York).
Hennig, H. and Rehorek, D. 1986. Static spectral sensitization of photocatalytic systems. In A.B.P. Lever (Ed.), Excited States and Reactive Intermediates, ACS Symposium Series, 307, 104-119.
Kaneko, M.,Takabayashi, N., Yamauchi, Y. and Yamada, A. 1984. Water photolysis by means of visible light with a system composed of Prussian blue and the tris(2,2'-biypridine) ruthenium (II) complex. Bull.Chem. Soc. Jpn., 57, 156-161.
Keggin, J.F. and Miles F.D. 1936. Structures and formulae of the Prussian blues and related compounds. Nature, 137, 577-578.
Ludi, A. and Güdel, H.L. 1973. Structural chemistry of polynuclear transition metal cyanides. Structure and Bonding, 14, 1-21.
Maer, K., Jr., Beasley, M.L., Collins, R.L. and Milligan, W.O. 1968. The structure of the titanium-iron cyanide complexes. J.Am.Chem.Soc., 90, 3201-3208.
Moser, J. and Grätzel, M. 1984. Photosensitized electron injection in colloidal semiconductors. J.Am.Chem.Soc., 106, 6557-6564.
Rasmussen, P.C. and Meyers, E.A. 1984. An investigation of Prussian blue analogues by Mössbauer spectroscopy and magnetic susceptibility. Polyhedron, 3, 183-190.
Robin, M.B. and Day, P. 1967. Mixed valence chemistry - survey and classification. Adv.Inorg.Chem.Radiochem., 10, 247-422.
Stalder, C. and Augustynski, J. 1979. Photoassisted oxidation of water at Be-doped polycrystalline TiO_2 electrodes. J.Electrochem.Soc., 126, 2007-2011.
Vlachopoulos, N., Liska, P., Augustynski, J. and Grätzel M. 1987. Very efficient visible light energy harvesting and conversion by spectral sensitization of high surface area polycrystalline titanium dioxide films. Submitted for publication.
Vrachnou, E., Vlachopoulos, N. and Grätzel, M. 1987. Efficient visible light sensitization of TiO_2 by surface complexation with $Fe(CN)_6^{4-}$. J.Chem.Soc.Chem.Commun., 868-870.

PHOTO-INDUCED ELECTRON TRANSFER REACTIONS IN POLYMER-BOUND

RUTHENIUM BIPYRIDYL COMPLEXES

Patricia M. Ennis and John M. Kelly
Chemistry Department, Trinity College, University of Dublin,
Dublin 2, Ireland.

ABSTRACT.

Photo-induced electron transfer between $[Ru(bpy)_3]^{2+}$- like centres covalently bound to positively-charged polymers (N-ethylated copolymers of vinylpyridine and $[Ru(bpy)_2(MVbpy)]^{2+}$) and viologens or Fe(III) has been studied using laser flash photolysis techniques. It is found that the backbone affects the rates of excited state quenching, the cage escape yield, and the back electron transfer rate because of both electrostatic and hydrophobic interactions. The effect of ionic strength on the reactions has been studied. Data on the electron transfer reactions of $[Ru(bpy)_3]^{2+}$ bound electrostatically or covalently to polystyrenesulphonate are also presented.

INTRODUCTION

Polymer-binding of photosensitisers and catalysts is a potentially useful approach for the construction of assemblies for the photodissociation of water by sunlight. (Kaneko and Yamada, 1984; Faulkner, 1984; Rabani and Sassoon, 1985). The macromolecule provides not only a backbone for the immobilisation and organisation of the components but can itself modify the efficiency of some of the steps. In this report we consider the role of the charge of the polymer on the efficiency of electron transfer from the photosensitiser excited state. Depending on the charge of the mediator, the charge on the polymer backbone might be expected to affect the rates of the excited state quenching, the extent of cage escape of the electron transfer products and their subsequent energy wasting back reactions. The species studied were ruthenium 2,2'-bipyridyl derivatives. Positively charged polymers were prepared by the N-ethylation of copolymers of 4-vinylpyridine and $[Ru(bpy)_2(MVbpy)^{2+}$ (MVbpy = 4-methyl-4'-vinyl-2,2'-bipyridyl) (Ennis et al 1986), while the negatively-charged compounds were based on polystyrenesulphonate. In order to assess the roles of charge and hydrophobicity both inorganic (Fe^{3+}) and organic [the positively charged methylviologen (MV^{2+}) and the neutral propylviologensulphonate (PVS)] electron acceptors were used. Pulsed dye laser (460 nm, 15 ns) excitation allowed the probing of the excited state luminescence or transient absorption and the formation and decay of the electron transfer products.

RESULTS

N-ethylated vinylpyridine copolymer.

The visible absorption and luminescence properties of the $[Ru(bpy)_3]^{2+}$-like centre in the N-ethylated polymer closely resemble those of $[Ru(bpy)_3]^{2+}$ and depend only slightly on the loading of the metal complex centre on the polymer (Ennis et al, 1986). The electron transfer reactions of a 1:11 copolymer have been studied in most detail.

Table 1 presents the data for the luminescence quenching of the copolymer excited state by methylviologen. It will be noted that at low ionic strengths the rate constant for the quenching is about an order of magnitude smaller than that for $[Ru(bpy)_3]^{2+}$, (Kalyanasundaram and Neumann Spallart, 1982), a feature which can be ascribed to the electrostatic repulsion of the positively-charged viologen by the polymer. Increasing the ionic strength of the solution is expected to neutralise the backbone charge, and in agreement with this prediction the rate of excited state quenching increases by an order of magnitude when the ionic strength is

TABLE 1: Emission lifetimes (in aerobic aqueous solution), Stern Volmer constants and rate constants for quenching by MV^{2+} of the excited $[Ru(bpy)_3]^{2+}$ - like centre of the N-ethylated vinylpyridine copolymer at various salt concentrations.

[NaCl]	τ ns	K_{SV} $dm^3 mol^{-1}$	k_q $dm^3 mol^{-1} s^{-1}$
0	420	15 [a]	3.5×10^7
0.066	420	38 [a]	9.0×10^7
0.155	420	67	1.6×10^8
0.37	440	115	2.6×10^8
0.70	460	144	3.1×10^8

[a] As Stern-Volmer plots non-linear these values are derived from the initial portion of the plot (i.e. up to $[MV^{2+}] = 3 \times 10^{-2}$ M.)

increased to 0.7 M with addition of NaCl. The extent of formation of the Ru(III) centre and the methylviologen radical cation (monitored by observation of the transient bleaching of the Ru(II) complex after flash photolysis in the presence of 0.1 M methylviologen) is somewhat less for the copolymer (0.17) than for $[Ru(bpy)_3]^{2+}$ (0.25). This result is not that expected from simple electrostatic considerations, where repulsion of the positively charged viologen radical by the backbone might be expected to enhance the cage escape yield, and may be an indication that other factors such as restricted diffusion due to polymer coiling or hydrophobic interactions are important. Both electrostatic and hydrophobic binding would be expected to reduce the rate of back reaction, and in agreement with this the rate constant for recombination of $MV^{\cdot +}$ and the Ru(III) centre on the copolymer is about one order of magnitude less than that for $[Ru(bpy)_3]^{3+}$ and the viologen radical (6 x 10^8 compared to 5 x $10^9 dm^3 mol^{-1}s^{-1}$).

Similar experiments were carried out with the zwitterionic viologen (PVS). It was found (in contrast to the behaviour of methylviologen) that the rate constant for quenching by PVS was only slightly reduced compared to that for $[Ru(bpy)_3]^{2+}$, and essentially independent of ionic strength (7 x 10^8 and 6.5 x 10^8 $dm^3 mol^{-1}s^{-1}$ in water and 1 M NaCl resp.) This contrasting behaviour is consistent with reduced electrostatic repulsion between the polymer and the quencher. The cage escape yield of reduced viologen from the polymer-bound Ru(III) was very small (<0.05), a feature consistent with both electrostatic and hydrophobic attraction of the $PVS^{\cdot -}$ for the polymer backbone.

In an attempt to examine the role of hydrophobic interactions, the electron transfer to ferric ion was studied in sulphuric acid solution (Table 2). It may be noted that the quenching rates in 0.007 M H_2SO_4 are similar for the copolymer and for $[Ru(bpy)_3]^{2+}$. At higher ionic strengths the rate of reaction is lower, probably a consequence of polymer coiling. The cage escape yield is only slightly lower for the polymer than for $[Ru(bpy)_3]^{2+}$, whereas the back reaction rate is actually increased for the polymer. This latter observation may be explained by assuming the formation of a ferrous-sulphate complex.

TABLE 2: Comparison of the rate constant k_q for quenching by Fe^{3+} of the excited state of the $[Ru(bpy)_3^{2+}]$ - like centre in the N-ethylated vinylpyridine copolymer or of $[Ru(bpy)_3]^{2+}$, the yield Φ for the formation of $[Ru(bpy)_3]^{3+}$ - centre and Fe^{2+}, and the rate constant (k_b) for the back electron transfer process.

Complex	k_q $dm^3mol^{-1}s^{-1}$	Φ (red)	k_b $dm^3mol^{-1}s^{-1}$
[H$_2$SO$_4$ = 0.007 M]			
Ru(bpy)$_3^{2+}$	2.5 x 10^9	1.0 ± 0.1	1.3 x 10^7
Copolymer	2.5 x 10^9	0.8 ± 0.1	1.2 x 10^8
[H$_2$SO$_4$ = 0.16 M]			
Ru(bpy)$_3^{2+}$	2.9 x 10^9	1.0 ± 0.1	8 x 10^6
Copolymer	1.3 x 10^9	0.9 ± 0.1	8 x 10^7

Polystyrenesulphonate polymers

Electron transfer reactions of $[Ru(bpy)_3]^{2+}$ are expected to be affected when the complex is bound to polystyrenesulphonate (PSSS). Such binding can be achieved by utilising the electrostatic attraction of the complex for the polymer or by preparing a copolymer of sodium styrenesulphonate.

The luminescence intensity of $[Ru(bpy)_3]^{2+}$ is increased in the presence of polystyrenesulphonate at concentrations of 10^{-4} M or greater. The intensity enhancement is greater in aerated (a factor of 1.8) than in degassed solution (a factor of 1.3), indicating that the lifetime of the complex when in contact with the polymer is affected both by its environment and by the reduced accessibility of oxygen. Scatchard analysis of the luminescence data as a function of PSSS concentration yields a binding constant of 8.5 x 10^{5}dm^{3}mol^{-1} and a binding number of 0.26., indicating a binding site containing between 3 and 4 styrenesulphonate units. Addition of NaCl to the solution reduces the luminescence enhancement as expected for elimination of the complex from the polymer.

The effect of PSSS on the oxidative quenching of $[Ru(bpy)_3]^{2+}$ excited state by MV^{2+}, PVS and Fe^{3+} has been investigated. For the cationic quenchers the extent of quenching was greatly enhanced, and static quenching is indicated. In both cases the reverse electron transfer must also be rapid, as electron transfer products having lifetimes greater than 50 ns could not be observed. With PVS dynamic quenching predominated, the rate of quenching in the presence of PSSS being somewhat lower ($8 \times 10^8 dm^3 mol^{-1} s^{-1}$) for [PSSS] = 1 mM than that in its absence ($1.3 \times 10^9 dm^3 mol^{-1} s^{-1}$). The cage escape yield of $PVS^{\cdot -}$ might be expected to be increased by repulsion away from the negatively charged polymer backbone. However the yield is less than that found in the absence of PSSS (0.06 for [PSSS] = 1.0 mM compared to 0.13 in its absence), possibly indicating that hydrophobic interactions restrain the electron transfer products from diffusing apart.

A polymer containing $[Ru(bpy)_3]^{2+}$-like centres bound has been prepared by copolymerisation of $[Ru(bpy)_2(MVbpy)]^{2+}$ and sodium styrenesulphonate (1:20 molar ratio) in the presence of β-cyclodextrin (O'Connell 1983). This sample is quenched by PVS with a rate constant ($1.0 \times 10^9 dm^3 mol^{-1}$) rather similar to that of the electrostatically-bound material. Interestingly the cage escape of products observed with this copolymer (0.18) is higher than that found for $[Ru(bpy)_3]^{2+}$. The reasons for this difference in reaction efficiency between the covalently- and electrostatically- bound materials are unclear at present and it is hoped to carry out a more extensive series of experiments with copolymer samples prepared by different procedures.

REFERENCES

Ennis, P.M., Kelly, J.M., and O'Connell, C.M., 1986, J.Chem.Soc., Dalton Trans., 2485.

Faulkner, R., 1984. Chem. Eng. News, 1984, 62(9), 28.

Kaneko, M and Yamada, A., 1984. Adv. Polym. Sci., 55, 1.

Rabani, J. and Sassoon, R.E., 1985, J. Photochem., 29, 7.

Kalyanasundaram, K. and Neumann-Spallart, M., 1982. Chem. Phys. Lett., 88, 7.

O'Connell, C.M., Ph.D. Thesis, University of Dublin, 1983.

Self Organization and Photofunctionalization of Supramolecular Systems *)

Photosensitive Polymeric Monolayers and Multilayers

L. Häußling, H. Ringsdorf

Institut für Organische Chemie
Becherweg 18-20
Johannes Gutenberg Universität
6500 MAINZ / Federal Republic of Germany

ABSTRACT

The design of light sensitive polymeric amphiphiles is discussed and their aggregation behaviour and the properties of LB- multilayers made from them are investigated. Enhanced thermal stability and annealing effects were found. In addition the photoreaction of amphiphilic Benzylammoniumsalts in Liposomes and LB- multilayers and its influence on the structure of these supramolecular systems was investigated.

1. INTRODUCTION

Order and Mobility are two basic principles of mother nature. The two extremes are realized in the perfect order of crystals with their lack of mobility and in the high mobility of liquids and their lack of order. Both properties are combined in liquid crystalline phases based on the selforganization of formanisotropic molecules. Their importance became more and more visible during the last years : in **Material science** they are a basis of new materials, in **Life science** they are important for many structure associated functions of biological systems. The main contribution of **Polymer science** to thermotropic and lyotropic liquid crystals as well as to biomembrane models consists in the fact that macromolecules can stabilize organized systems and at the same time retain mobility. The synthesis, structure, properties and photofunctionalization of polymeric amphiphiles in monolayers and multilayers will be discussed.

---

*) This is a prolonged abstract of a paper given at the 2nd workshop on Photochemical & Photobiological Processes for Producing Energy-rich Compounds, September 22nd-25th 1987, in Carmona/Spain.

2. MONO- AND MULTILAYERS FROM PREPOLYMERIZED AMPHIPHILES

The route to polymeric membranes was originally based on the polymerization or polycondensation of preoriented monomeric amphiphiles (Bader et al.,1985). The direct use of polymers is based on a certain mobility of the system allowing the free organization of the orientable side groups partially independent of the polymeric main chain. This can either be reached via low degrees of polymerization or by the introduction of mobility increasing units into the polymer. One of the possibilities is the incorporation of side chain or main chain spacers into these polymers, thereby introducing mobility and microphaseseparation. This so called spacer concept (Ringsdorf et al., 1975), that could also be successfully applied to polymeric liquid crystals (Finkelmann et al.,1978), postulates a partial decoupling of the ordering function of the side chains from the stabilizing function of the polymeric main chain. The introduction of the main chain spacer could be achieved by copolymerization of a polymerizable lipid together with a hydrophilic comonomer (Laschewsky et al.,1986)

Combining the properties of polymers as well as amphiphiles, these copolymers exhibit some interesting features : The ease of LB -multilayer formation and annealing effects of these LB- multilayers as demonstrated in fig.5. In addition recent results showed that for copolymers with fluorocarbon or mesogenic side groups the order of multilayer is restored on cooling even after complete loss of the layer correlation at high temperatures (Erdelen, 1987 ; Ringsdorf et al.,1987).

3. Results and Discussion

The variation of structure and function of organized systems can easily be achieved by photochemical means (Ramamurthy, 1986). There is a strong interdependence between organized systems and the photoreactions carried out in these systems. On the one hand, the result of the photoreaction has a strong influence on the ordering of the systems. On the other hand, the outcome of almost any photochemical reaction is altered in organized systems (Whitten, 1979).

3.1. Photoreactions of the Benzylammoniumsalts

Such an alteration can be demonstrated with the photocleavage of meta disubstituted Benzylammoniumsalts. They give upon irradiation in their long wavelength UV- absorption band, a homolytical and a heterolytical cleavage of the benzylcarbon nitrogen bond.

The heterolytically formed product results from solvolysis of the benzylcation, whereas the product of the homolytic cleavage is the corresponding toluene, produced by radical disproportionation via fast hydrogen abstraction from the aminylium radical (Ratcliff, 1971; McKenna, 1980; Haubs 1987).

If this photoreaction is carried out in organized systems, all cage processes (ion- and radical recombination and radical disproportionation) are favoured through the increased lifetime of these cages. Correspondingly, quantitative analysis and comparison of the photoproducts in isotropic solution and organized systems shows a strongly different behaviour. Thus the photolysis of 3,5-dihexadecyloxybenzyl- N,N,N-triethyl-ammoniumbromide in liposomal solution gave a considerably higher yield of 3,5-dihexadecyloxytoluene than the corresponding photoreaction of 3,5-dimethoxybenzyl-N,N,N-triethylammoniumbromide in isotropic solution (water) as depicted in table 1.

Table 1: Yields of Photolysis of long chain (3,5-dihexadecyloxybenzyl-N,N,N-triethylammoniumbromide) and short chain (3,5-dimethoxybenzyl-N,N,N-triethylammoniumbromide) Benzylammoniumcompounds in water.

Product :	isotropic solution water	organized media liposomes
Toluene	74 %	88 %
Benzyl-alcohol	26 %	12 %

3.2. Amphiphilic quarternary Benzylammoniumsalts

To study the photoreaction of monomeric and in monolayers and amphiphiles in polymeric LB- multilayers, the following Benzylammoniumcompounds were synthesized :

Radical polymerization yielded the corresponding homopolymers. Polymeric amphiphiles with main chain spacers were prepared via copolymerization with 2-Hydroxyethylacrylate.

The aggregation behaviour of the photosensitive monomeric and polymeric amphiphiles at the gas - water interface was studied at different temperatures. The effect of polymerization on the phase behaviour of the Benzylammoniumamphiphiles in the monolayer is given in Fig. 1. The effect of the introduction of a main chain spacer as compared to the homopolymer is demonstrated in Fig. 2.

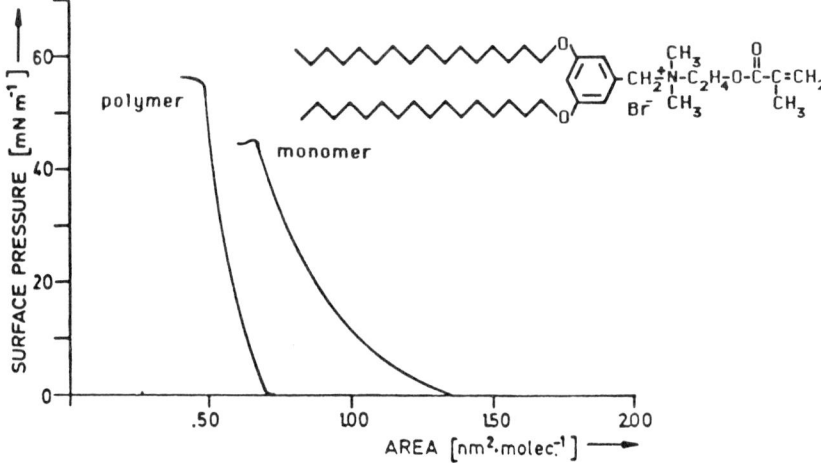

Fig. 1 Pressure - Area diagrams of monolayers of monomeric and prepolymerized Benzylammoniumamphiphiles at the gas - water interface

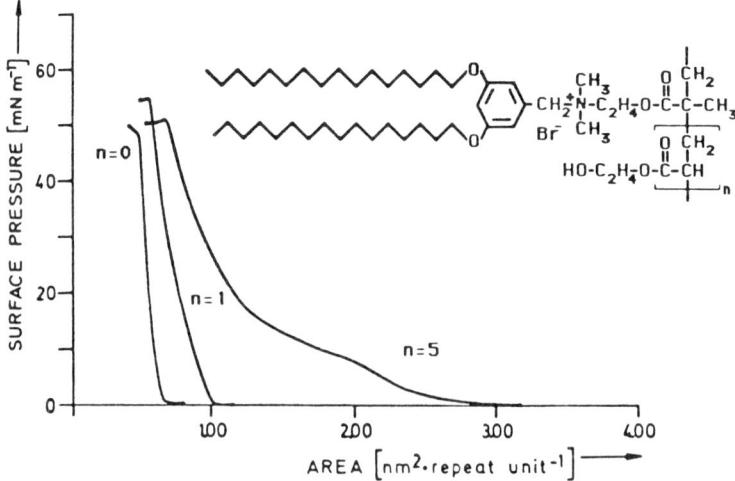

Fig. 2 Pressure - Area diagrams of monolayers of these prepolymerized Benzyl-ammoniumamphiphiles at the gas - water interface. Effect of the introduction of a hydrophilic main chain spacer.

As shown in the Figure 2, compared to the corresponding homopolymer, the copolymers show a "liquid expanded phase" depending on the comonomer content. This liquid expanded phase behaviour is not caused by a phase transition in the side chain as found with many natural and synthetic lipids. In this case it can be explained as an entropy driven coiling (**B**) - uncoiling (**A**) process of the copolymeric main chain (Frey et al., 1987).

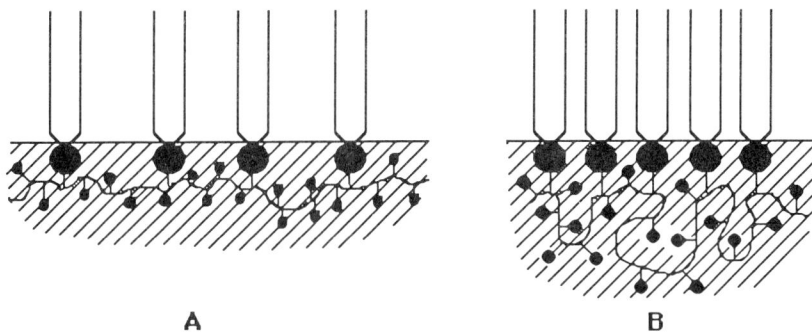

A   B

3.3. LB- Multilayers from prepolymerized Benzylammoniumamphiphiles

LB- multilayers of these polymers were built up and their thermal and photochemical behaviour was investigated. The increase of absorbance following each successive deposition of two bilayers on a hydrophobized quartz support is shown in the following figure :

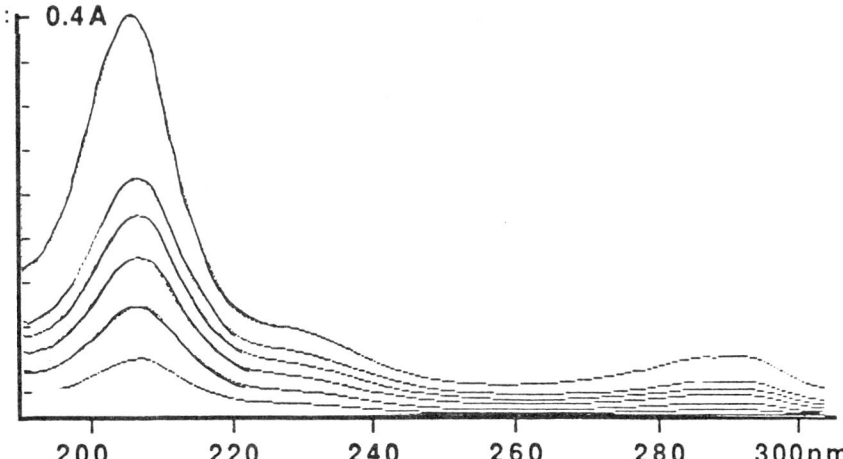

Fig. 3 UV-spectra of LB-multilayers of prepolymerized Benzylammoniumamphiphiles after transfer of 4,8,12,16,20 and 40 monolayers on a hydrophobized quartz support

It was also possible to transfer monolayers of these prepolymerized Benzylammoniumamphiphiles on a Polyethylenetherephthalte (PET) - support. These multilayers could be studied by Small Angle X-ray Scattering (SAXS) experiments. The scattering diagram is shown in the following figure :

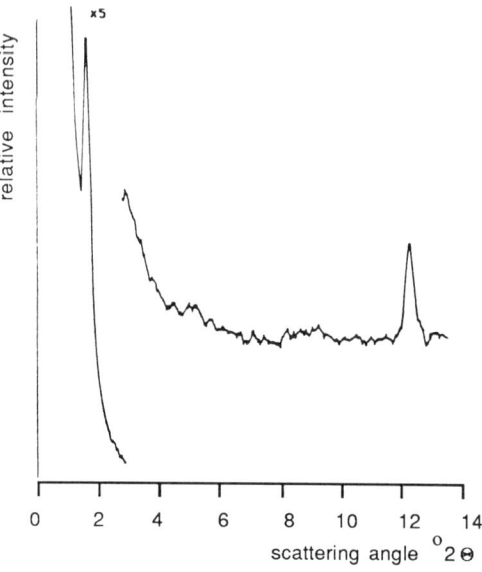

Fig. 4 SAXS-diagram of 20 monolayers of prepolymerized Benzylammoniumamphiphiles on a PET-support. The peak at 12.35 °2Θ is caused by the support.

Only a first order scattering peak at 1.7 °2Θ was found, corresponding to a layer spacing of 53 Å and pointing to a tilt angle of the alkyl chains of about 35 °. An interesting annealing effect of these LB- multilayers could be observed during the temperature dependent SAXS studies, as shown in Fig. 5. The layer correlation increased considerably at 50 °C. However, upon further heating to 85 °C, irreversible loss of the layered stucture was observed.

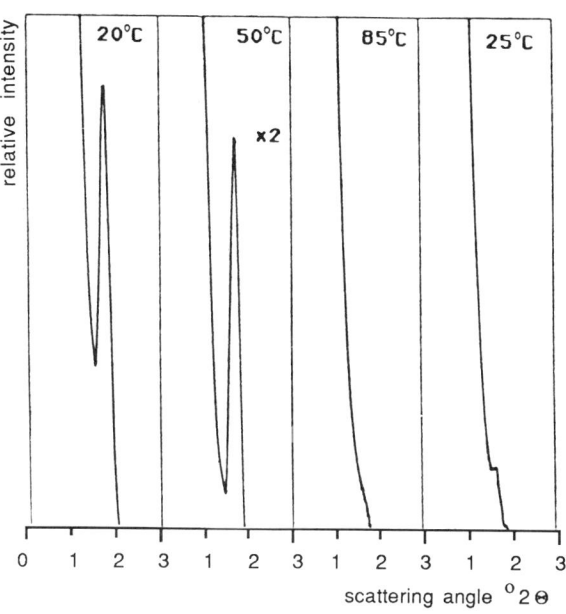

Fig.5 SAXS-diagrams of first order peaks of Fig. 4 at 20,50,85°C and after consecutive cooling to 20 °C.

3.4. Photoreaction of Benzylammoniumsalts in LB -Multilayers

The photoreaction of the Benzylammoniumsalts in LB- Multilayers was first investigated by UV- spectroscopy :

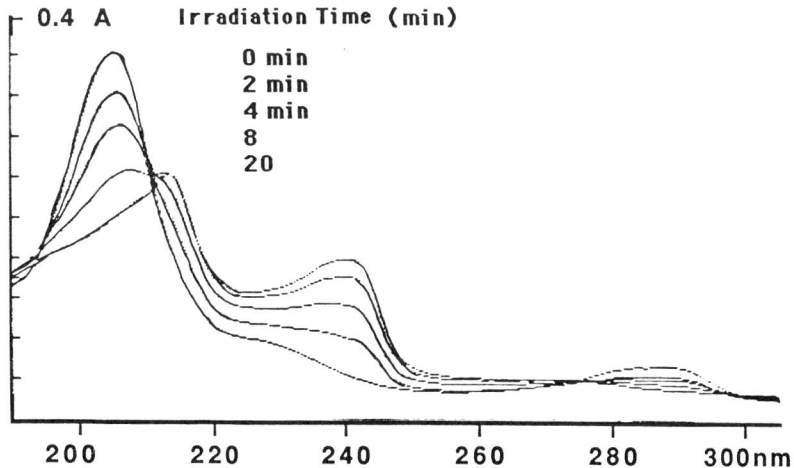

Fig. 6 UV-spectra of 40 monolayers of prepolymerized Benzylammoniumamphiphiles on a hydrophobized quartz support after 2,4,8 and 20 min irradiation time (Osram HBO 100)

The photoreaction shows an isosbestic point only during the first few minutes of irradiation. This is due to a changing microenvironment of the chromophore during the photoreaction, an indication of structural changes caused by this process.

To investigate these structural changes, irradiation time dependent SAXS experiments were carried out :

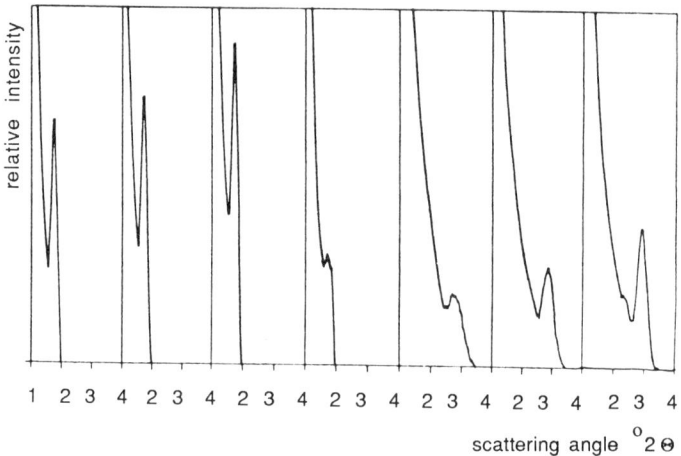

Fig.7 SAXS-diagrams of the first order peaks of Fig. 4 after irradiation times of 15 sec.,1,3,5,8,20 min.

It was interesting to find that after the total collapse of the original layer structure of the polymeric LB- Multilayers after 3 min irradiation time, a new scattering peak was found after continued irradiation. This points to the generation of a new layer structure of the hydrophobic alkylfragments in between the independent ionic polymeric layers.

Wether the new layer spacing of 30 Å corresponds to a non oriented layer or - as indicated by the formal layer distance - to an interdigitated oriented double layer can not be decided by these experiments. A speculative view of the ongoing photoprocess is summarized in the following three figures (A,B,C):

LB- Multilayers After Short Irradiation (<30 sec)

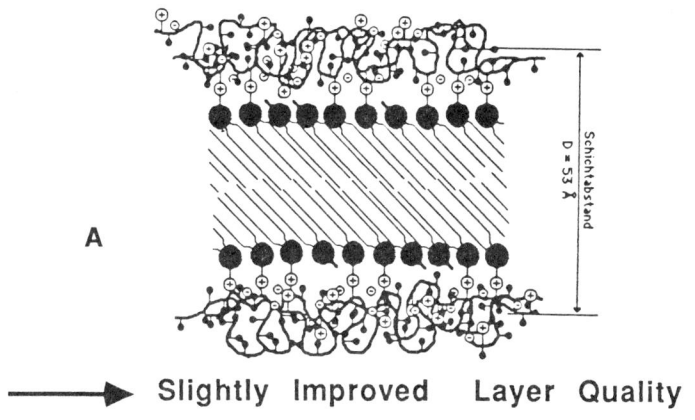

A

→ Slightly Improved Layer Quality

After 3 min Irradiation Time

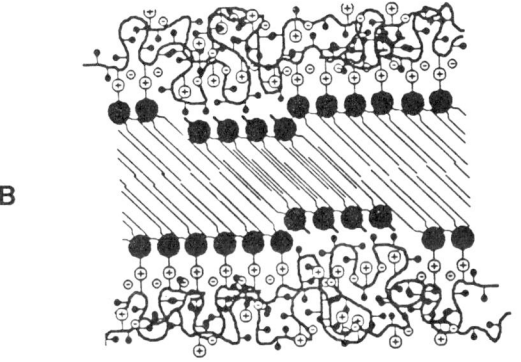

B

After Completion of the Photoreaction

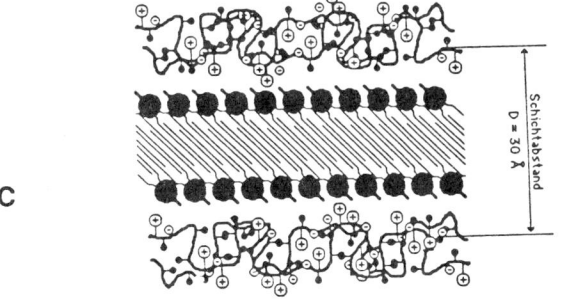

C

→ Probably Generation of a New Layer Structure

REFERENCES

Bader, H., Dorn, K., Hupfer, B., Ringsdorf, H. 1985. Polymeric Monolayers and Liposomes as Models for Biomembranes. Adv. Polym. Sci. 64 , 1

Erdelen, C. 1987. Untersuchungen an orientierten Systemen aus monomeren und polymeren Amphiphilen mit perfluorierten Seitenketten. Diplomarbeit Mainz 1987

Finkelmann, H., Koldehoff, J., Ringsdorf, H. 1979. Synthesis and Characterization of Liquid-Crystalline Polymers with Cholesteric Phase. Angew. Chem. Int. Ed. Engl. 17, 935

Frey, W.,Ringsdorf, H., Sackmann, E., Schneider, J. 1987. Preparation, Microstructure and thermodynamic Properties of Homogeneous and Heterogeneous Compound Monolayers of Polymerized and Monomeric Surfactants on the Air/Water Interface and on Solid Substrates. Macromolecules 20 , 1312

Haubs, M., Ringsdorf, H. 1987. Photosensitive Monolayers, Bilayer Membranes and Polymers. Nouv. J. Chim. 11 , 151

Laschewsky, A., Ringsdorf, H., Schneider, J. 1986. Oriented Supramolecular Systems - Polymeric Monolayers and Multilayers from Prepolymerized Amphiphiles. Die Angew. Makromol. Chem. 145/146 ,1

McKenna, J., et al. 1980. The Mechanism of Photolysis of some Benzyltrimethylammonium Salts in Water and Alcohols. J. Chem. Soc. Perkin II 1980 , 77

Ramamurthy, V. 1986. Organic Photochemistry in Organized Media. Tetrahedron 42 , 5753

Ratcliff, M.A. and Kochi, J.K., 1971. Solvolytic and Radical Processes in the Photolysis of Benzylammoniumsalts. J. Org. Chem. 36 , 3112

Ringsdorf, H. 1975. Structure and Properties of Pharmacologically Active Polymers J. Polym. Sci. Symp. 51 , 135

Ringsdorf, H., Schneider, J., Schuster, A. 1987. Langmuir-Blodgett Multilayers from Copolymers with Amphiphilic and Mesogenic Side-Groups. Abstracts 3rd Int. Conf. LB films Göttingen 1987. 282

Roviello, A., Sirigu, A. 1975.Mesophasic Structures in Polymers. A Preliminary Account on the Mesophases of some Polyalkanoates of p,p´-dihydroxy- α,α - dimethyl benzalazine. J. Polym. Sci. Polym. Lett. Ed. 13 455

Whitten, D.G. 1979. Photochemische Reaktionen oberflächenaktiver Moleküle in Systemen monomolekularer Schichten - Steuerung der Reaktivität durch die Umgebung. Angew. Chem. 91 , 472

PHOTOGENERATION OF HYDROGEN: THE PHOTOCHEMICAL WAY
OF STORING SOLAR ENERGY

G. Munuera, A. Fernández and J.P Espinós
Instituto de Ciencias de Materiales. Centro mixto
CSIC-Univ. Sevilla. P.O.Box 1115. Sevilla. SPAIN

ABSTRACT

Water cleavage induced by uv-irradiation of aqueous suspensions of M/TiO_2 polycrystalline samples (M=Pt,Rh) has been studied. Only H_2 evolution was observed while oxygen could not be detected during the experiments. The presence of O_2 really suppress H_2 photo-generation while oxygen photo-uptake is observed. A mechanism is proposed for the photoinduced water splitting that involves generation of hydrogen and hydrogen peroxide, this latter being retained on the TiO_2 and probably corroding the TiO_2 particles, without appreciable loss of photo-activity.

INTRODUCTION

Hydrogen photogeneration from H_2O is a potential method for solar energy conversion (Harriman and West, 1982) having several advantages, such as a direct way of storage, cheaper transport, a well-known technology to take profit of it and a potential use in combustion cells for it direct conversion "in situ" in electrical power with a high yield.

Previous work using band gap irradiated metal loaded semiconductors has shown that hydrogen is really evolved from aqueous suspensions of these systems though O_2 evolution was not observed in most cases (Duonghong et al., 1981; Mills et al., 1982).

This paper shows some experiments using Rh/TiO_2 and Pt/TiO_2 which suggest that oxygen must be incorporated to the TiO_2 support as peroxo-species in a considerable amount.

EXPERIMENTAL

P-25 anatase TiO_2 ($S_{BET}=49\pm1\ m^2g^{-1}$) from Degussa was purified by a thermal oxidation/rehydration treatment described elsewhere (Munuera et al., 1979).

The Rh/TiO_2 sample (2.5% by weight) was prepared by incipient wetness impregnation with $RhCl_3$ and then reduced under H_2 flow at 423K for 2 h. Pt/TiO_2 (0.06% by weight) was suplied by J.Kiwi and has been prepared by ionic interchange (Kiwi et al., 1984) on the same support. The experiences where carried out at ca. 313K in a pyrex flask similar to that described by Grätzel et al. (Yesodharan et al.,1983) using 25ml of a 1M NaOH solution.

Aqueous suspensions (ca. 50 mg of sample) were irradiated with a 200W Osram HBO bulb with a total energy output at the flat window of the flask of 470 mW/cm^2. Analysis of H_2 and O_2 evolved into the flask free volume (ca. 17.6 cm^3) was made by GC using a molecular sieve 5Å column and Ar as carrier gas. Detection limit was 1µmol for O_2 and 0.1 µmol for H_2 in that volume.

Prior to each irradiation the suspension was deaereated with Argon until N_2 and/or O_2 peaks where not observed. In some of the experiments measured doses of O_2 and/or H_2 or H_2O_2 were injected.

Analysis of H_2O_2 in the aqueous phase were carried out at the end of each experiment by the peroxidase method with a sensibility better than 10^{-5}M.

RESULTS

When suspensions of the M/TiO_2 (M=Pt,Rh) polycrystalline samples in 1M NaOH are irradiated in the uv ($\lambda \leqslant 400$ nm) hydrogen is generated, Fig. 1(a), showing a initial induction period and a slow decay (S-shaped curve) while O_2 was not detected throughout the whole experiments (ca. 9h). Similar experiments using the TiO_2 support did not show evolution of either H_2 or O_2 into the gas phase. Enzimatic test for H_2O_2 after each experiment was negative, indicating that H_2O_2, if formed, was not released to the aqueous phase. If before irradiation 1 ml$_{NPT}$ of O_2 is injected in the flask, Fig. 1(b), equilibrium is ready reached in the dark while irradiation leads now to inmediate O_2 photo-uptake H_2 evolution being partially hindered. The rate of oxygen photo-uptake was rather similar for both samples and similar to that observed under the same conditions on the TiO_2 support, what suggest that it takes place mainly on the oxide. Enzimatic test for H_2O_2 was also negative after these experiments. It is worthy of note that the rates of H_2 evolution, in the absence of oxygen, are of the same order to that of O_2 photo-uptake when it is present in the gas phase, so that competitive reactions seems to occurs according to:

$$2H^+ + 2e^- \longrightarrow H_2$$

$$O_{2(g)} + 2e^- \longrightarrow O_{2(ad)}^{2-} \quad (\text{or } O_2H^-)$$

Successive experiments, Fig. 1(c), on the same sample give slightly higher rates for H_2 evolution, that now occur without initial induction period. This may be related to the reduction of the metallic particles in the previous experiments, since it vanishes when samples are pre-treated with

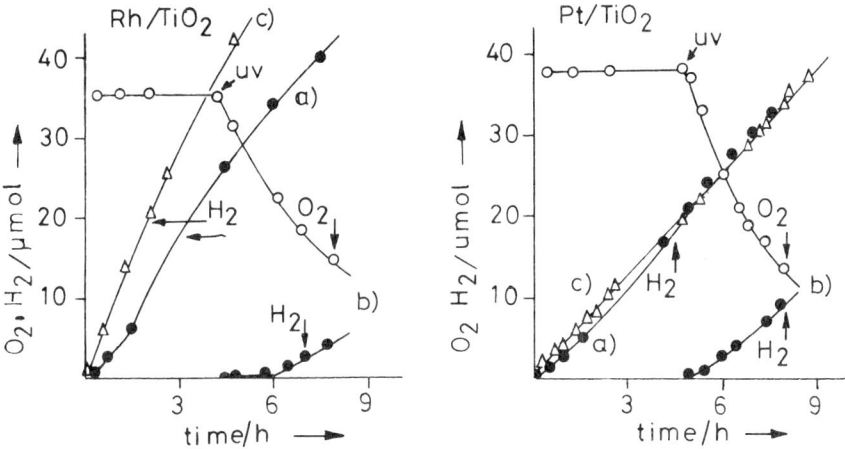

Fig.1.- Gas phase evolution during the uv-irradiation of M/TiO_2 suspensions (1M NaOH): a) original sample; b) with O_2 in the gas phase; c) a fresh sample pre-reduced in H_2 at 298K.

H_2 at 298K before irradiation. The new rates of H_2 evolution remains almost unchanged during several set of experiments (each 9h irradiation) on both samples, while O_2 was not detected.

Since O_2 was not evolved during irradiation together with H_2 and it becomes photo-adsorbed according to reaction (2), while the enzimatic test shows the absence of H_2O_2 in the liquid phase, the study of H_2O_2 interaction with both TiO_2 and M/TiO_2 (M=Pt, Rh) was examined under the same experimental conditions used in the irradiation experiment to generate H_2. A saturation coverage of ca. 1 H_2O_2/nm^2 was obtained with TiO_2 at concentrations of H_2O_2 in the aqueous phase $\leqslant 10^{-2}$M in good agreement with previous results (Munuera et al., 1980) while only a very slow decompositon was observed. However, when H_2O_2 was injected into the M/TiO_2 suspensions a fast evolution of O_2 was observed in the dark, Fig. 2, until total decomposition of the H_2O_2 occurs, thus indicating an important catalytic effect of the metallic particles, which is very high for the Rh/TiO_2 sample. These results suggest that while H_2O_2 can be adsorbed under the basic conditions on the TiO_2 support, where it remains stabilized probably as HO_2^-, H_2O_2 in the aqueous phase would inmediately decompose even in the dark giving O_2. According to our results in Fig. 1(b), this oxygen should be adsorbed again under uv-irradiation on the support.

Provided that H_2-O_2 recombination is a process that should be related

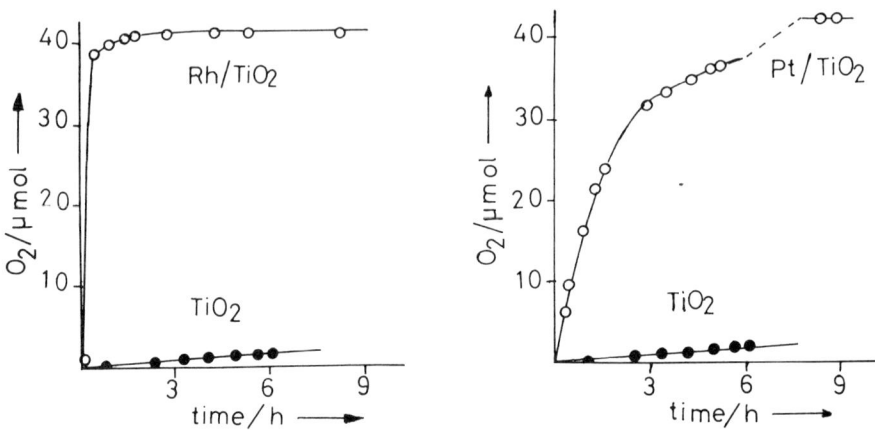

Fig.2.- Evolution of O_2 from H_2O_2 in dark on Rh/TiO_2 and Pt/TiO_2.

with the actual yield of water cleavage, Fig.3 shows the results of several recombination experiments on M/TiO_2 (M=Pt, Rh). For both samples a decrease in the initial pressure of H_2 and O_2 could be recorded in the dark. The relative rates close 2:1, indicates stoichiometric recombination to give H_2O. The reaction does not occur on TiO_2 during the same period of time, so we must conclude that it takes place on the metallic particles, Pt being much more efficient than Rh in this process.

As shown in Fig.3, when the flask was irradiated in the presence of

Fig.3.- Stoichiometric H_2-O_2 recombination in M/TiO_2 suspensions (1M NaOH) at ca. 313K in the dark.

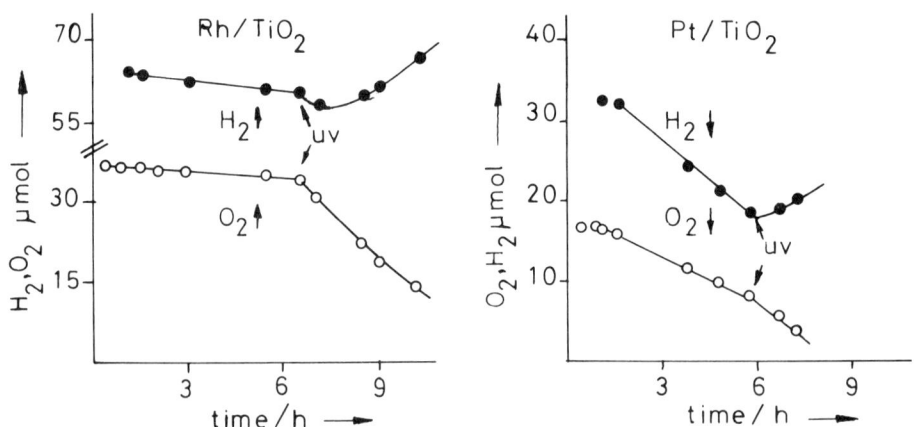

the H_2-O_2 mixture, O_2 photo-uptake was observed while H_2 is photogenerated at increasing rates as far as O_2 becomes exhausted from the gas phase. A similar experiment with TiO_2 shows O_2 photoadsorption while only a small decrease in H_2 pressure can be detected after 4 h of irradiation.

DISCUSSION

In most previous work to produce water photocleavage on metal loaded semiconductors (i.e. TiO_2, $SrTiO_3$, CdS, etc.) an stoichiometric H_2/O_2 evolution according to:

$$H_2O \xrightarrow[M/TiO_2]{h\nu} H_2 + 1/2 O_2 \quad (1)$$

has not been observed. However, from our previous work with Rh/TiO_2 (Munuera et al., 1984) the alternative reaction

$$2H_2O \xrightarrow[M/TiO_2]{h\nu} H_2 + H_2O_2 \quad (2)$$

was proposed which may be extended now to Pt/TiO_2 samples.

The driving force for the processes being the generation of asymmetric barriers between the liquid-TiO_2 and metal-TiO_2 interfaces, as shown in Fig.4, which allows charge carriers to splitt and then to react with OH^- and H^+ species.

However, a simple calculation assuming a rate for H_2O_2 generation similar to that for H_2 evolution in Fig. 1(c), gives amounts of H_2O_2 of ca. 10 molecules/nm^2, after 9 h of irradiation, which corresponds to 8-10 times the monolayer capacity for these species on the TiO_2 support in both M/TiO_2

Fig.4.- Maximum biass gain in photoelectrolysis

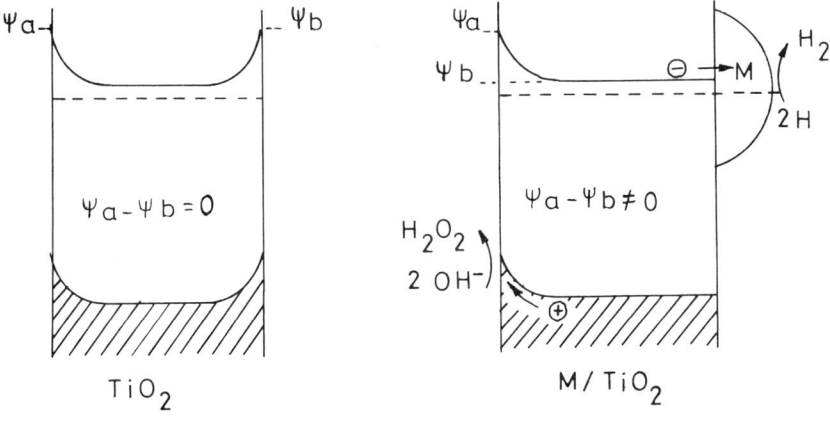

samples. If only one monolayer (ca. 1 H_2O_2/nm^2) remains on the TiO_2 support most of the H_2O_2 would be released to the liquid phase were they will give a ca. 10^{-3} M solution, a concentration that is about a hundred times higher than the detection limit of our enzimatic test which however, was always negative. This result can be explained taking into account our data in Fig. 2, that shows a complete and fast decomposition of H_2O_2 in solution, catalysed by the metallic particles even in the dark, to give O_2 in the gas phase. However oxygen produced in this way should be photoabsorbed, according to our data in Fig. 1(b), on the TiO_2 support under uv-irradiation to form again O_2^{2-} (or O_2H^-) species, this leading to recombination of holes and electrons and eventually to stop the generation of hydrogen, what really was not observed during our experiments. Therefore, we must conclude that during the irradiation process a considerable amount of peroxo-species would accumulate at the TiO_2 support where it is likely that the photogenerated H_2O_2, would react with the hydroxilated TiO_2 surface according to:

$$OH^-_{ad} + H_2O_2 \longrightarrow H_2O + O_2H^-_{ad} \quad (3)$$

forming a thin layer of hydroxy-peroxo-compounds involving several monolayers of the TiO_2 particles (TiO_2 corrosion). It is worthy of note that the observed sustained photo-activity of the samples only will occur if the modified TiO_2 surface layers, containing the peroxo-species, have rather similar band gap absorption characteristics than the TiO_2 so that irradiation with $\lambda \leq 400$nm will still produce the same charge carriers as for TiO_2.

According to our model the nature of the barrier existing at this liquid-TiO_2 modified surface must be determined at each time by the amount of charged species existing at the irradiated surface, and in particular by the O_2^{n-}/O_2H^- concentrations. In principle, these species must produce a Shottky barrier that should by important for charge carrier splitting under uv-irradiation, while under H_2 atmosphere ohmic contact should be formed at the metal-TiO_2 interface (Aspnes et al., 1983) thus allowing an easy and fast electron transfer to protons at the metallic particles, as schematically shown in Fig. 4.

It is likely that the Shottky barrier at the liquid-TiO_2 modified surface will increase with the amounts of peroxo-species so that the rate of oxygen photo-uptake would progressively decrease while the energy and rate of the holes reaching the surface will increase so that at some stage O_2 photogeneration may start according to:

$$2O_2H^- + 2h^+ \longrightarrow H_2O_2 + O_2 \quad (4)$$

once the rate of this latter process become higher than that of O_2 photo-uptake. In fact, after a long period of irradiation a steady state would be reached in which stoichiometric $H_2:O_2$ evolution could be observed. Our model seems to explain why several authors (Duonghong et al., 1981) observe a long delay for stoichiometric O_2 evolution, while others (Mills et al., 1982) observe O_2 evolution in lower amount than the stoichiometric value, that only will occur when the rate of reaction (4) become higher than that O_2 photo-uptake. A situation that was not reached in our experimental conditions.

Finally it should be stressed that H_2-O_2 recombination in our samples is a slow process, particularly with Rh/TiO_2, so that a good yield of H_2 (and O_2) can be expected with this sample even in a close system as the one used in our work.

Since the proposed mechanism for water cleavage on M/TiO_2 (M=Rh, Pt) assumes that generation of peroxo-compounds on the TiO_2 is important in water cleavage by these type of systems, the study of these compounds seems of paramount interest in the understanding of the whole process.

Acknowledgment

Authors thank the CAICYT and Fundación R.Areces for the finantial support.

REFERENCES

Aspnes, D.E. and Heller, A. 1982. Photoelectrochemical Hydrogen Evolution and Water-Photolyzing Semiconductor Suspensions: Properties of Platinum Group Metal Catalyst-Semiconductor Contacts in Air and in Hydrogen. J. Phys. Chem., 87, 4919-1929.
Duonghong, D., Borgarello, E. and Grätzel, M. 1981. Dynamics of Light-Induced Water Cleavage in Colloidal Systems. J. Am. Chem. Soc., 103, 4685-4690.
Harriman, A. and West, M.A., eds. 1982. "Photogeneration of Hydrogen". Academic Press, London.
Kiwi, J. and Grätzel, M. 1984. Optimization of Conditions for Photochemical Water Cleavage: Aqueous Pt/TiO_2 (Anatase) Dispersions under U.V. Light. J. Phys. Chem., 88, 1302-1307.
Mills, A. and Porter, G. 1982. Photosensitised Dissociation of Water using Dispersed Suspensions of n-type Semiconductors. J.C.S.Faraday Trans.I, 78, 3659-3669.
Munuera, G., Rives-Arnau, V. and Saucedo, A. 1979. Photo-adsorption and Photo-desorption of Oxygen on Highly Hydroxilated TiO_2 SurfacesI.J.C.S. Faraday Trans. I, 75, 736-746.
Munuera, G., R. González-Elipe, A., Soria, J. anf Sanz, J. 1980. Photo-adsorption and Photo-desorption of Oxygen on Highly Hydroxilated TiO_2 Surfaces III. J.C.S. Faraday Trans. I, 76, 1535-1546.
Munuera, G., Soria, J., Conesa, J.C., R. González-Elipe, A., Navío, A, López-Molina, E.J., Muñoz, A., Fernández, A. anf Espinós, J.P. 1984.

Catalysis on the Energy Scene (Ed. S.Kaliaguine and A.Mahay). Elsevier, Amsterdam. pp 335-346.

Yesodharan, E. and Grätzel, M. 1983. Photodecomposition of Liquid Water with TiO_2-Supported Noble Metal Clusters. Helv. Chim. Act., 66, 2145-2153.

INORGANIC PHOTOSYNTHESIS: THE PHOTOFIXATION OF THE ATMOSPHERIC DINOTROGEN ON TRANSITION METAL OXIDES

R.I.Bickley[*] and J.A.Navio-Santos[**]

[*]Schools of Studies in Chemistry and Chemical Technology
University of Bradford. Bradford BD7 1DP (UK)
[**]Instituto de Ciencias de Materiales-CSIC/Dpto. de Química Inorganica.Facultad de Químicas.Universidad de Sevilla. Sevilla (Spain).

ABSTRACT

A concise review of data relative to the photofixation of dinitrogen on transition metal oxides is presented. Analysis of data is focussed upon the nature and potential efectivenes of a process which can occur in nature on the surface of particular inorganic materials when exposed to direct sunlight. The significance of this process is discussed. Finally, some new and selected results are presented which outline particular mechanistic aspects of the photoassisted nitrogen process.

INTRODUCTION

1.1.- Literature

The current phase of interest in dinitrogen fixation started in the early 1960's when bacterial nitrogenase was first extracted in active form (Carnahan et al.,1960) and the chemistry of dinitrogen fixation received a further boost in 1965 when the first stable and well-characterized dinitrogen complexes $(Ru(NH_3)_5(N_2))^{2+}$ were isolated (Allen et al., 1965).

The route for which the atmospheric dinitrogen is fixed in the soil is a complex process which requires the concourse of some soil microorganism which allow the fixation in extremely mild conditions. This possibility contrasts with the high energy requirement of the Haber-Bosch process, in which the fixation of dinitrogen is brought about by the union of nitrogen and hydrogen to form ammonia. Although, the ammonia route is apparently energetically the least satisfactory, it is currently and probably will remain for many years the only one which is commercially viable for industrial dinitrogen fixation.

In addition to the classical Haber process, recent efforts have been directed to finding an alternative economic process for fixing N_2 to meet - increasing demands for nitrogeneous fertilizers and others industrial requirements. Thus, numerous results of the recent advances in the Chemistry of Nitrogen Fixation via transition metal complexes have reported (Chatt et al.

1978; "A treatise on dinitrogen fixation", Hardy ed.1979; "New trends in the chemistry of nitrogen fixation", Chatt ed.1980).

Nevertheless, perhaps the most important developments during the last decade has been the discovery of a photoassisted solid-catalyzed reduction - of dinitrogen by water, involving illumination of semiconductors materials (mainly TiO_2 and Fe/TiO_2) to produce ammonia and hydrazine to the photoreducing reactions summarised as:

1) $N_2^g + 3 H_2O^l \xrightarrow[TiO_2]{h\nu \ (\lambda \geq 365 \ nm)} 2 NH_3^g + 3/2 \ O_2^g \quad \Delta G^{298} = + 765 \ kJ/mol$

2) $N_2^g + 2 H_2O^l \xrightarrow[TiO_2]{h\nu \ (\lambda \geq 365 \ nm)} N_2H_4^l + O_2^g \quad \Delta G^{298} = + 625.7 \ kJ/mol$

The same authors (Schrauzer et al. 1977) have reported that iron doping enhances the photocatalytic reactivity of TiO_2 (rutile) for photochemical ammonia synthesis from N_2 and H_2O. This system has been extensively explored by others authors (Augugliaro et al. 1982; Schiavello et al. 1984) using photochemical reactors. Likewise, the presence of adsorbed ammonia has been detected by Navio and Bickley (1984) after the reduction of dinitrogen with irradiated substochiometric $TiO_{2(1-x)}$. Recently, Endoh et al. (1986) have reported that dinitrogen can be photosynthetically reduced to ammonia with irradiated substochiometric tungsten oxides WO_{3-x} and with tungsten trioxide WO_3. The main conclusion of this investigation (Endoh et al. 1986) was that the ammonia photoproduction in the gas/solid system occurred photocatalytically with turnovers greater than unity. On the contrary, the reaction in aqueous solution occurred, mainly, by a thermal path, rather than by a photocatalytic one, according to the following equation:

3) $WO_{2.96} + 0.04 \ H_2O + (0.04/3) \ N_2 \longrightarrow WO_3 + (0.08/3) \ NH_3$

This photogeneration of ammonia in the gas phase was interpreted by the authors (Endoh et al. 1986) as follow: firstly, H_2O molecules adsorbed on WO_{3-x} are split into hydrogen and oxygen under UV-irradiation, followed by reaction of the dinitrogen with hydrogen to produce ammonia.

Khader et al. (1987) performed the photosplitting of water using a catalyst consisting of a mixture of $\sim$ 90% α-Fe_2O_3 and $\sim$ 10% Fe_3O_4. Using this catalyst the same authors (Khader et al. 1987) have reported the photoassisted reduction of dinitrogen with water.

From the reported results it is evident that dinitrogen photoreduction and water phosplitting seems to be parallel catalytic processes.

Recent work by Lichtin (Lichtin et al. 1984, 1986) have reported the possibility to photoreduce the N_2 by liquid water, using a wide range of transition metal oxides and others inorganic materials which are catalytically active for the process.

In addition to reduction, oxidations reactions of adsorbed dinitrogen seems to be promoted on irradiated TiO_2 surfaces, despite the reaction being thermodynamic not very unfavorable.

4) $\quad N_2^g + O_2^g \xrightarrow[TiO_2^s]{h\nu} 2\ NO^g \quad (\Delta G^{298} = +\ 181,4\ kJ/mol)$

(data from Schrauzer et. al 1986)

In this context, Bickley et al. (1979) have reported that surfaces of TiO_2 which contain pre-adsorbed hydrogen peroxide molecules, have a capacity to fix $\overline{N_2}$(gas) in the dark, generating oxidised nitrogen species.

1.2.-Significance of the research

Because titanium is abundant in the earth's crust, the photocatalytic reactions for fixing the atmospheric dinitrogen must been considered as a part of the nitrogen cycle. Thus, in a recent work (Schrauzer et al. (1986), have reported the possibility to photoreduce and/or photo-oxidizethe the dinitrogen on TiO_2 and other inorganic materials containig titanium, under similar conditions to those operating in nature.

On the other hand, the results reported by Bickley (Bickley et al. 1979) claim the possibility of a photo-chemical process for the photo-oxidative dinitrogen fixation on TiO_2 and materials related, which could occur spontaneouslly in nature, because of the evidence (Voltz et al.1972,1976; Munuera et al. 1980; Bickley et al. 1986; Munuera et al. 1986 and Graetzel et al. 1987) for the photogeneration of hydrogen peroxide on irradiated TiO_2 surfaces in the presence of oxygen and water, the other two major components of the Atmosphere.

Conservative estimates by Schrauzer (Schrauzer et al. 1986)suggest that annually about 10 millions tons of ammonia are generated in the semi--arid regions of the earth. This implies that a 10% of the available nitrogen in the biosphere appears to originate from solar driven electrochemical fixation of atmospheric dinitrogen via naturally occurring semiconducting minerals.

Special conclusions must be drawn from the process discussed by Lichtin et al. (1986) from energetically downhill reactions, when an adsorbed organic molecule, such as cellulose, is oxidized, while photogenerated electrons reduce dinitrogen and protons to ammonia. Indeed, the discovery of such kinds of process is of great interest, from the agriculture point of view, especially if it is assumed that in soils the semiconducting minerals are coated with an adsorbed organic layer.

The reaction of dinitrogen with irradiated inorganic materials has been also tested for reactions different from the photofixation of dinitrogen. In this way, Lichtin et al. (1984) have reported that a mixture of N_2 and CO_2 exposed on irradiated transition metal oxides of the VIII group produced a substantial number of amino-acids someones of which are considered essential in the biochemistry of humans.

1.3.-Some mechanistics considerations

From results reported in the literature it is concluded that a certain number of doubts exist about the mechanism for the photochemical fixation of molecular nitrogen. For example, in a similar experiment to those reported by Bickley (Bickley et al.1979), Schrauzer (1986) have detected the formation of traces of nitrite, but not nitrate, from aqueous extracts of TiO_2 after exposure in air under UV-irradiation. This result, shows that nitrogen photo-oxidation proceeds even with low efficiency. But, as TiO_2 promotes the photo-oxidation of ammonia by oxygen (Ritchie et al.1965; Mozzanega, 1975), nitrogen oxidation products could be also generated by a route which suggests initialy, the photogeneration of ammonia and subsequently its photo-oxidation, such as was observed recently by Navio (Bickley et al.1986). In contrast with this result it can be mentioned that because the photo-oxidative fixation of dinitrogen seems to be more feasible, from the thermodynamic point of view, the photofixation of dinitrogen to ammonia may occur through some oxidative pathway implicaiting an intermediate oxidative dinitrogen species (NO_2^-, NO_3^-, $N_2O_2^=$, etc) which can been subsequently photoreduced to ammonia.

On the basis of the results it is inferred that evidence exists for fixing atmospheric dinitrogen via inorganic materials. The mechanism remains to be established and in this sense, research on dinitrogen complexes of transition metals is important, because these N_2-complexes can provide "models" for understanding photochemical or biological nitrogen fixing systems.

EARLY RESULTS AND THEIR INTERPRETATION

Experimental

Photochemical experiments were performed in a greasless glass high vacuum system of an ultimate vacuum of ca. 10^{-4} Pa. An schematic diagram of apparatus is shows in Fig.1. Powdered specimens of TiO_2 ($\sim$ 1 g) were contained in a thermostatted quartz reaction cell. Irradiation of the surface on the powdered specimens was effected from above by 500W medium pressure mercury arc contained within a water-cooled pyrex glass condenser. Pressures of pure gases were measured either with a calibrated Pirani gauge or with a calibrated pressure transducer, the latter being used also for gaseous mixtures. Analysis of the strongly adsorbed surface species were made using temperature programmed desorption (t.p.d.) linked to contonuous monitoring of the gas phase with a mass-spectrometer(Vacuum Generator, Micromass 2A) in its scanning mode of operation.

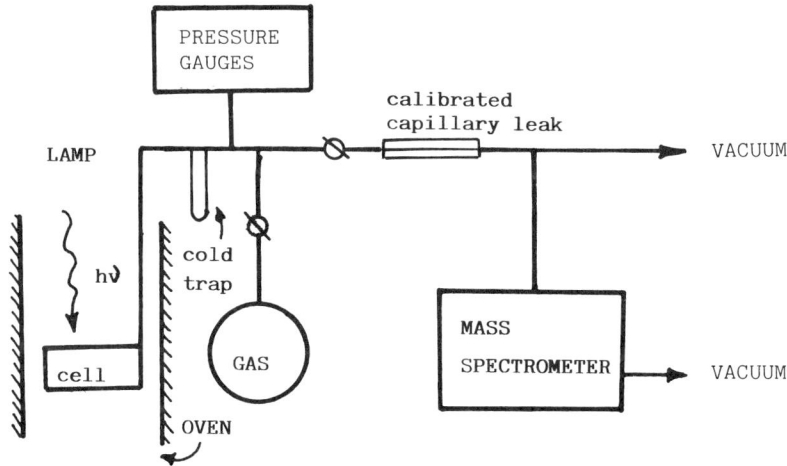

Fig.1.- Schematic Diagram of Apparatus

Pure TiO_2 rutile specimens (supplied by Tioxide International Ltd.; ex sulphate of 4.2 m^2g^{-1}) were evacuated at 550ºC for 12 h., followed by a slow cooling to room temperature. This treatment, led to a substoichiometric specimen $TiO_{2(1-x)}$ ($x \approx 1.5 \times 10^{-6}$). The interaction with this surface, of a vapour pressure of water at room temperature, and the subsequent evacuation of the gas phase for 5 h., leads to a surface situation on the specimen, hereafter as " reduced/hydroxylated" $OH/TiO_{2(1-y)}$.

Results and Discussion

From results reported in the literature, the possibility of fixing the atmospheric dinitrogen until ammonia using TiO_2 as well as other materials containing titania, under solar irradiation, is clearly evident. But the - question is raises how is this process occurs in the oxidising conditions which exist in our present atmosphere. In order to try to elucidate to the above - question, the following experiment have been performed.

A first experiment was performed in order to check the ability of pure TiO_2 (rutile) for the photofixation of dinitrogen on simulated natural conditions. Fig. 2 shows the t.p.d. profiles of pure TiO_2 which has been exposed to a mixture N_2/O_2 gas (air) under UV-irradiation. From Fig. 2 it is clear that the photofixation of dinitrogen occurs on pure TiO_2 (in the presence of air) generating oxidised nitrogen species which desorbed as NO, in agreement with previous results (see Bickley et al. 1979). This results can be interpreted on the basis of previous results as follow: N_2 is fixed oxidatively in presence of hydrogen peroxide (Bickley et al. 1979), and oxygen photoadsorption on hydroxylated TiO_2 occurs with the formation of adsorbed hydrogen peroxide at room temperature (Bickley et al. 1986). Both processes are conected (see Bickley and Navio, 1987).

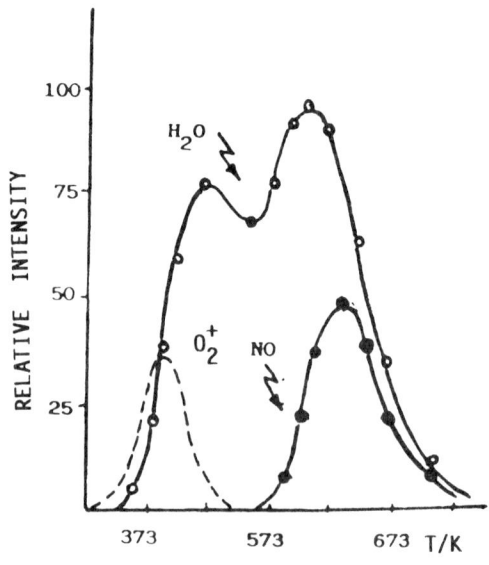

Fig.2.- T.p.d. of TiO_2(hydroxylated) after exposure to UV-irradiation(5 h) in the presence of N_2/O_2(air) at 300K

In a second experiment a sample OH/TiO$_{2(1-y)}$ was exposed during 5 h. under UV-irradiation in the presence of pure oxygen ($\sim$ 1 torr). Mass-spectrometric analysis of the gas phase, after the photoexperiment, showed only oxygen. Temperature programmed desorption from this TiO$_2$ specimen, after irradiation clearly shows a profile of water and oxygen which compared with previous results (Bickley et al. 1986) offers evidence for the photogeneration of H$_2$O$_2$ (ads) under these conditions. This results can be interpreted assuming that the evolution of hydrogen observed from this "reduced/hydroxylated" surfaces, irradiated in vacuo (Bickley and Navio, 1984) is totaly suppressed by the presence of oxygen, involving thus, the photogeneration of hydrogen peroxide capable of giving some active oxygen form for fixing dinitrogen, according with the proposed mechanism I.

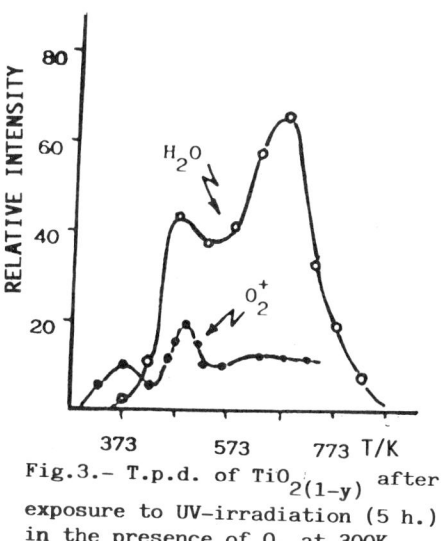

Fig.3.- T.p.d. of TiO$_{2(1-y)}$ after exposure to UV-irradiation (5 h.) in the presence of O$_2$ at 300K

MECHANISM I

$$\cdot H_{ads} + O_{2\,gas} \longrightarrow \cdot HO_{2\,ads}$$

$$\cdot HO_{2\,ads} + \cdot HO_{2\,ads} \longrightarrow H_2O_{2\,ads} + O_{2\,gas}$$

$$H_2O_{2\,ads} + Ti^{+3} \longrightarrow H_2O_{ads} + O^*_{ads} + Ti^{+4} \text{ (where } O^* \text{ could be the specie } O^- \text{)}$$

In a third experiment, the same surface OH|TiO$_{2(1-y)}$ was exposed for 5 h. under UV-irradiation to an equimolecular mixture of N$_2$/O$_2$ ($\sim$ 1 torr). No gaseouss hydrogen was found during the photoexperiment. T.p.d. from the surface, after the photoexperiment (Fig.4) revealed that nitrogen is fixed on the surface in two forms, reductively as ammonia (NH$_3$) and oxidatively as (NO$_x^{-z}$), the fixed nitrogen being desorbed, mainly as dinitrogen. These results raises mechanistics questions about alternative possibilities for the photofixation of dinitrogen:

i) preferentialy, via a reductive step to form ammonia, which is subsequently photo-oxidized to (NO_x^{-z}) by means of the hydrogen peroxide photogenerated, according to scheme II, route (i).

ii) preferentialy, via an oxidative step to form NO_x^{-z} and subsequently, by photoreduction to NH_3 by means of hydrogen photogenerated radicals, according with scheme II, route (ii).

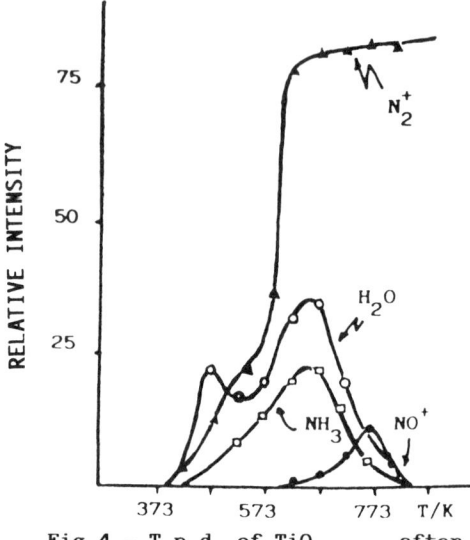

Fig.4.- T.p.d. of $TiO_{2(1-y)}$ after exposure to UV-irradiation (5 h.) in the presence of N_2/O_2 at 300K

MECHANISTIC SCHEME II

(see Bickley and Navio, 1984)

$$N_2 \text{ gas} \xrightarrow[\text{OH}/TiO_{2(1-y)}]{h\nu/UV} NH_3 \text{ ads}$$

$$O_2 \text{ gas} \xrightarrow[\text{OH}/TiO_{2(1-y)}]{h\nu/UV} H_2O_2 \text{ ads}$$

$$NH_3 \text{ ads} + H_2O_2 \text{ ads} \longrightarrow (NO_x^{-z}) \xrightarrow{t.p.d.} NO \text{ gas} \qquad \text{ROUTE (i)}$$
$$\downarrow$$
$$NH_2OH_{ads} (?) \longrightarrow N_2O_{ads} \xrightarrow{t.p.d.} N_2 \text{ gas} + O_{ads}$$

Hydrogen photogenerated radicals + **Oxidised** nitrogen species
$$\downarrow$$
Adsorbed ammonia   ROUTE (ii)

ACKNOWLEDGEMENTS

The authors gratefully acknowledge the finantial support of the British Council and the Spanish Minitry of Education through an Anglo-Spanish "Acción Integrada. In addition, one of us (Navio-Santos) is grateful to "Junta de Andalucia" for the finantial support through "Estancias breves en el extranjero" during the period 1984/85 and 1985/86. Both authors, wishes to thank also Professor G.Munuera for his encouragement during the course of this work.

REFERENCES

Allen, A.D. and Sernoff, C.V. 1965
    J. Chem. Comm. 621,
Augugliaro, V., Lanricella, A., Rizutti, L. Shiavello, M. and Sclafani, A. 1982.
    Int. J. Hydrogen Energy, 7, pp. 845-849.
Auglugiaro, V. D'Alba, F. Rizutti, L., Schiavello, M. and Sclafani, A. 1982
    Int. J. Hydrogen Energy, 7, pp. 851-855.
Bickley, R.I., and Vishwanatham, V., 1979
    Nature, 280, pp 306-307
Bickley, R.I., Navio, J.A., Schiavello, M., Rizutti, L., and Yue, P.L.,1984
    Proc. 8th. Int. Congress on Catalysis. Verlag Chimie, Basel, Vol.III
    pp 383-394
Bickley, R.I., Navio, J.A., Jayanty, R.K.M., and Vishwanatham, V., 1986
    "Homogeneous and Heterogeneous Photocatalysis"
    Pelizetti, E. and Serpone, N. (Eds.) by D.Reidel Publishing Company
    NATO-Series , pp 555-565.
Bickley, R.I. , Navio, J.A., Vishwanatham, V. 1986
    Proc. 6th. International Conference on Photochemical Conversion and
    Storage of Solar Energy. Paris (July, 1986)
Carnahan, J.E. Mortense, L.E., Mouer, H.F. and Castle, J.E. 1960
    Biochem, Biophys. Acta. 38, pp. 188.
Chatt, J. Dilworth, J.R., and Richards, R.L. (1978)
    "Recent Advances in the Chemistry of Nitrogen Fixation".
    Chem. Reviews, Vol. 78, N°6, pp. 589-625.
Chatt, J. (Editor). 1980.
    "New Trends in the Chemistry of Nitrogen Fixation ".
    Academic Press. London.
Endoh, E., Leland, J.K. and Bard, A.J. 1986.
    J. Phys. Chem. 90, pp. 6223-6226
Gräetzel, M., and Kiwi, J. 1987.
    J. Mol. Cat. 39, pp. 63.
Hardy, R.W.F. (General Editor) 1979.
    " A Treatise on Dinitrogen Fixation".
    A Wiley- Interscience Publication. New York.
Khader, M.M., Lichtin, N.N.; Vurens, G.H., Salmeron M. and Somorjai, G.A. 1987.
    Langmuir, 3, pp. 303-304.
Lichtin,N.N. 1984
    USA Patent 4427509
Lichtin,N.N. and Vijayakumar, K.M. 1984
    USA Patent 4427510
Lichtin, N.N. and Barman, E. 1984
    USA Patent, 4443311
Lichtin, N.N. and Vijayakumar, K.M. 1986
    J.Indian Chem. Soc., 63, pp 29-34
Lichtin, N.N. and Vijayakumar, K.M. 1986
    USA Patent 4612096
Mozzanega, H. 1975
    Thesis CNAM, Lyon (France)
Munuera, G., Gonzalez-Elipe, A.R., Sanz, J. and Soria, J. 1980
    JCS, Faraday I, 76, pp 1535
Munuera, G., Gonzalez-Elipe, A.R., Espinos, J.P. and Navio, J.A. 1986
    J.Molecular Structure, 143, pp 227-230
    Previous evidence was presented by Munuera, G. and Navio, J.A. at

IV Reunion Nacional de los Grupos de Trabajo Relacionados con la Investigacion en el Campo de la Adsorcion. Sevilla (Spain)-September 1979.

Ritchie, M. and McLean, W.R. 1965
    J.Appl. Chem. $\underline{15}$, pp 452-460

Schiavello, M. and others 1984
    Proc. 8th. Int. Congress on Catalysis. Verlag Chimie. Basel. Vol. III, pp 383- 394.

Schrauzer, G.N. and Guth, T.D. 1977
    J.Am. Chem. Soc. $\underline{99}$, pp 7189-7193

Schrauzer, G.N., Guth, T.D. , Salehi, J., Strampachi, N., Nan Hung, L. and Palmer, M.R. 1986
    "Homogeneous and Heterogeneous Photocatalysis"
    Pelizetti, E. and Serpone, N. (Eds.) by D.Reidel Publishing Company pp 509

Voltz, H.G., Kamp, J., Fitzky, H.G. 1972 y 1976
    Farbe um Lack, $\underline{75}$, pp 736 (1972)
    Farbe um Lack, $\underline{82}$, pp 805 (1976)

RECENT TRENDS IN THE SEARCH FOR NEW PHOTOSENSITIZERS

A. Juris and V. Balzani

Dipartimento Chimico "G. Ciamician" dell'Universita' di Bologna
I-40126 Bologna, Italy, and
Istituto FRAE-CNR, via Castagnoli 1, I-40126 Bologna, Italy

ABSTRACT

Recent trends in the search for new photosensitizers to be used for solar energy conversion processes are illustrated. Particular emphasis is given on the Ru(II)-polypyridine family, the use of new cyclometalated ligands, and the use of cage-type ligands.

INTRODUCTION

It seems likely that artificial devices for the photochemical conversion and storage of solar energy will be complex systems, constituted by several molecular species arranged so as to achieve a long-lived charge separation (Balzani, 1987). In this frame, studies on photoinduced electron transfer reactions in homogeneous solution may appear to have a quite limited role. Nevertheless, studies on homogeneous systems are important because they can contribute to elucidate fundamental principles, and to design, check and optimize new photosensitizers, donors and acceptors that will eventually be used as building blocks for supramolecular devices (Balzani et. al., 1986).

The most important component of any solar energy conversion device is, of course, the photosensitizer. In this paper we will try to illustrate briefly our current studies on the search for new photosensitizers based on transition metal complexes.

An ideal photosensitizer must satisfy several stringent requirements (Balzani et. al., 1986): 1) stability towards thermal and photochemical decomposition reactions; 2) sufficiently intense absorption bands in a suitable spectral region; 3) high efficiency of population of the reactive excited state; 4) long lifetime in the reactive excited state; 5) suitable ground state and excited state potentials; 6) reversible redox behavior; 7) good kinetic factors for outer sphere electron transfer reactions.

A survey of the chemical compounds shows immediately that entire classes of molecules can be discarded as potential photosensitizers. For various reasons, the most promising field to explore in the search for photosensitizers is that of transition metal complexes.

THE Ru(II)-POLYPYRIDINE FAMILY

The most widely used photosensitizer is the Ru(bpy)$_3^{2+}$ complex (bpy = 2,2'-bipyridine) (Kalyanasundaram, 1982; Watts, 1983; Juris et. al., in press). It exhibits an absorption band at 450 nm with ε = 14000, the efficiency of conversion from the excited state originally populated by excitation and the reactive state (η_{isc}) is unity, the lifetime of the reactive excited state is in the µs range, the excited state is oxidized at -0.86 V, and the reduction potential of the oxidized form of the complex is +1.26 V. The excited state energy is 2.12 eV, the self exchange rate for electron transfer is larger than 10^6 $M^{-1}s^{-1}$.

The main drawbacks of Ru(bpy)$_3^{2+}$ are as follows: 1) relatively small absorption in the visible region (the absorption spectrum of Ru(bpy)$_3^{2+}$ does not match well the emission spectrum of the sun); 2) a relatively high threshold energy (with the lowest excited state at 2.12 eV, the maximum thermodynamic efficiency for solar energy conversion is 21%; 3) a small turnover number (reported values range from 10^2 to 10^3); and 4) a high cost.

In the last few years, one strategy to find better photosensitizers has been to explore the potentiality of the Ru-polypyridine family. More than 300 luminescent Ru(II)-polypyridine complexes have been reported in the literature, involving about 200 different ligands (Juris et. al., in press). This huge amount of work has shown that, in principle, one can remedy the first two drawbacks by a suitable choice of ligands; in particular, by replacing one or two bpy with a ligand which is easier to reduce such as biq (2,2'-biquinoline), one can obtain a much better absorption spectrum because of the presence of both Ru--->bpy and, at lower energy, Ru--->biq CT bands, with a noticeable increase in solar harvesting, and also with a lower threshold energy (1.73 eV instead of 2.12 eV) which allows a much larger thermodynamic efficiency of solar energy conversion (from 21% to 29%, Juris et. al., 1982).

Let now us consider possible remedies to the small turnover number. One reason for the low turnover number is the occurrence of a ligand photodissociation reaction which takes place from the lowest ^{3}MC excited state after an activated surface crossing from the luminescent ^{3}MLCT state (Kalyanasundaram, 1982; Watts, 1983; Juris et. al., in press). To

avoid this decomposition reaction, two strategies can be followed. One is
to increase the energy difference between the ^{3}MLCT and the ^{3}MC states,
and the other is to link together the ligands so as to have a cryptate-
type ligand that prevents dissociation. To modify the ^{3}MLCT-^{3}MC energy
gap one can either shift to lower energy the ^{3}MLCT state with a more
easily reducible ligand as biq (see before), or increase the energy of
the ^{3}MC state using metals or ligands with higher ligand field strength.
Replacing Ru(II) with Os(II) increases the energy gap between ^{3}MLCT and
^{3}MC, but the lifetime of the ^{3}MLCT is usually much shorter (Kober et.
al., 1986). A promising class of ligands with high ligand field strength
appears to be that of the cyclometalating ligands, which are structurally
similar to the bipyridine ligands but have a higher ligand field strength
because the carbanion donor is much higher in the spectrochemical series
that the corresponding nitrogen donor. However, the first results along
this direction are somewhat disappointing. The complex Ru(bpy)$_2$L$^+$, where
L = 2-(3-nitrophenyl)pyridine, exhibits better absorption properties than
Ru(bpy)$_3$$^{2+}$ and, as was expected, it is photoinert (Reveco et. al., 1986).
Unfortunately it only emits at 77 K and not above 140 K, which indicates
that its excited state lifetime at room temperature has to be extremely
short. Anyway, the use of the cyclometalating ligands has opened a new
avenue towards the design of coordination compounds with long-lived
luminescent excited states. Some dozens of these compounds have already
been prepared with Rh, Pd, Ir, and Pt as metals (Maestri et. al., 1987;
Ohsawa et. al., 1987). They exhibit long lifetimes and beautiful emission
spectra at low temperature, and there is again an extraordinary possibility to tune their photophysical properties because of an interplay between LC and CT excited states. Together with some excellent properties,
however, these orthometalated complexes exhibit several drawbacks (e.g.,
very poor absorption in the visible region). Further research in this
area is needed in order to understand whether this class of complexes may
provide good photosensitizers.

THE Co(sep)$^{3+}$ COMPLEX

As mentioned above, an alternative way to prevent photodissociation
(with consequent increase in the turnover number) is that of linking
together the ligands so as to have a cage-type metal complex. The first
example of this strategy in another field was given some years ago by
Sargeson, who has synthetized the Co(sep)$^{3+}$ complex (Figure 1), i.e. the

cage analog of the $Co(NH_3)_6^{3+}$ complex (Creaser et. al., 1977). In principle, linking together the ligands so as to make a cage does not modify the spectroscopic properties and redox potentials, but prevents the decomposition of the complex. This $Co(sep)^{3+}$ complex has the same spectroscopic and redox properties of $Co(NH_3)_6^{3+}$, but it does not undergo dissociation upon reduction or light excitation (Pina et. al., 1985a).

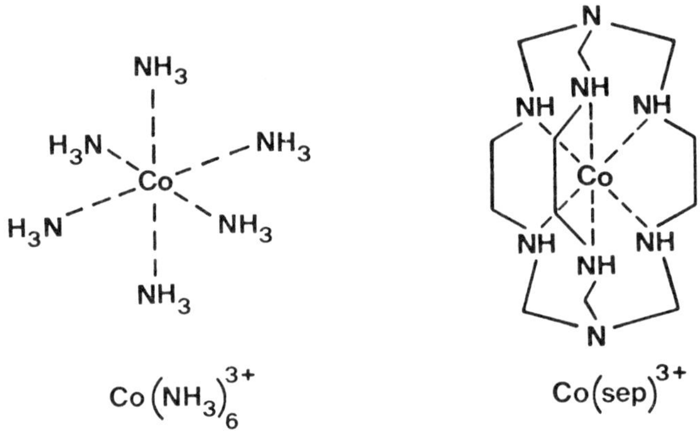

Fig. 1. Structural formulae of $Co(NH_3)_6^{3+}$ and $Co(sep)^{3+}$.

These two complexes also exhibit a quite different photochemical behavior when they are paired with anions. Both of them give rise to ion pairs (IP) with I^- or $C_2O_4^{2-}$ ions, as shown by the appearance of new absorption bands in the near UV spectral region. Excitation of the $Co(NH_3)_6^{3+}-I^-$ ion pair in the ion pair charge transfer (IPCT) band causes the transfer of an electron from the anion, which is in the second coordination sphere, to the Co(III) center which is reduced to Co(II). The Co(II) complexes are known to be labile and indeed the ligand dissociation process is fast enough to compete with the back electron transfer, giving rise to a photoredox decomposition reaction with high quantum yield. In the analogous ion-pair system involving $Co(sep)^{3+}$, light excitation in the IPCT band leads again to I radicals and to a Co(II) complex. In this case, however, the Co(II) complex cannot undergo ligand dissociation because the ligand is a cage. So, no net reaction takes place. However, when the photoreaction is carried out in the presence of oxygen, $Co(sep)^{2+}$ can be oxidized back by oxygen and I_2 can be extracted from the solution, for example with $CHCl_3$. In this system, $Co(sep)^{3+}$

plays the role of an electron transfer photosensitizer, since the overall reaction is oxidation of I^- by O_2 (Pina et. al., 1985a).

A similar but, perhaps, more interesting case is that of the ion pairs between $Co(sep)^{3+}$ and oxalate ions. Excitation of the ion pair in the IPCT band leads to the formation of the Co(II) complex and of an oxalate radical which undergoes a fast decomposition reaction. Thus, the back electron transfer reaction is prevented and $Co(sep)^{2+}$, which is a good reductant, can accumulate in the system. When colloidal platinum is present in the solution the $Co(sep)^{2+}$ complex reduces H^+ to H_2, with regeneration of the $Co(sep)^{3+}$ complex. In this system, the $Co(sep)^{3+}$ complex plays again the role of an electron transfer photosensitizer (Pina et. al., 1985b).

In conclusion, whereas $Co(NH_3)_6^{3+}$ is useless because it undergoes a fast photodecomposition reaction, the analogous $Co(sep)^{3+}$ complex may be employed as an electron transfer photosensitizer because of its intrinsic stability in the excited state and in the reduced form. In the same way, one can think to use cage-type polypyridine ligands for Ru complexes, so as to prevent ligand dissociation reactions.

POLYPYRIDINE CAGE-TYPE LIGANDS

The Eu^{3+} and Tb^{3+} complexes of the cage-type ligands shown in Figure 2 (Rodriguez-Ubis et. al., 1984) have already been prepared by Lehn and

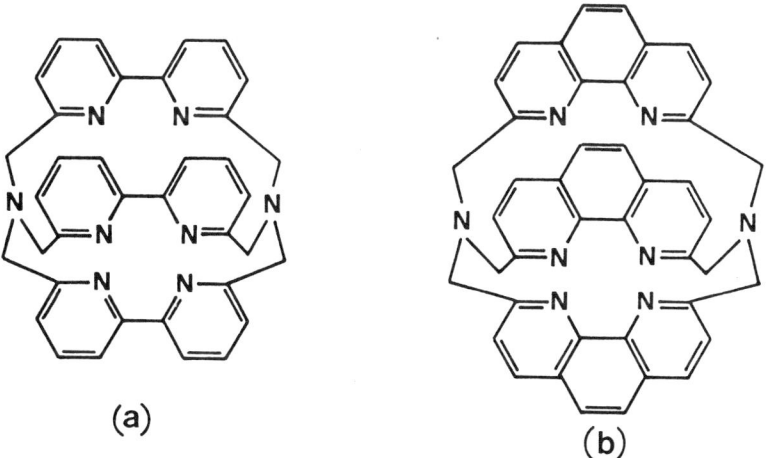

Fig. 2. Macrobicyclic ligands containing 2,2'-bipyridine (a) or 1,10-phenanthroline (b).

co-workers, but the Ru(II) complex has not yet been prepared and there might also be problems concerning the right shape and size of the cage. If the Ru-N bonds were too long or the bite angles were unfavorable, the ligand field strength would be small and the decrease in energy of the lowest ^{3}MC excited state would compromise the lifetime of the ^{3}MLCT excited state, as it happens for the Ru(trpy)$_2^{2+}$ complex (trpy = 2,2',2"-terpyridine). A more promising cage-type polypyridine ligand is that recently synthetized (Grammenudi and Voegtle, 1986). This cage ligand is much more flexible than that prepared by Lehn and can better fit small ions. The Fe(II) complex of this ligand has already been prepared (Grammenudi and Voegtle, 1986) and we are now trying to sinthetize the Ru(II) complex.

CONCLUSIONS

In conclusion, some suggestions for the design of a suitable photosensitizer are as follows:
- to choose compounds that contain a transition metal ion, in order to ensure the presence of a redox site at a suitable potential;
- to prefer metal ions of the II or III transition row, in order to avoid that the lowest excited state is MC in nature, which would imply extremely short lifetime and photochemical and redox lability;
- to select ligands which are easy to reduce, in order to have a MLCT excited state as the lowest excited state; this assures long lifetime, small S-T energy gap, high η_{isc}, intense absorption bands at low energy, and favorable electronic factor for redox quenching;
- to link together the ligands so as to encapsulate the metal ion, in order to prevent thermal and photochemical decomposition.

Of course, there is no reason why the Ru-polypyridine complexes have to be the best photosensitizers. Studies should be performed on other families of transition metal compounds, with emphasis on complexes of inorganic ligands and on binuclear and polynuclear species.

REFERENCES

Balzani, V.; Juris, A.; Scandola, F. 1986. In "Homogeneous and heterogeneous photocatalysis", E. Pelizzetti and N. Serpone (Eds.), Reidel (Dordrecht), p. 1.
Balzani, V. (Ed.) 1987. "Supramolecular photochemistry", Reidel (Dordrecht).

Creaser, I.I.; Harrowfield, J.M.; Hertl, A.J.; Sargeson, A.M.; Springborg, J.; Geue, R.J.; Snow, M.R. 1977. J. Am. Chem. Soc., 99, 3181.
Grammenudi, S.; Voegtle, F. 1986. Angew. Chem. Int. Ed. Engl., 25, 1122.
Kalyanasundaram, K. 1982. Coord. Chem. Rev., 46, 159.
Kober, E.M.; Caspar, J.V.; Lumpkin, R.S.; Meyer, T.J. 1986. J. Phys. Chem., 90, 3722.
Juris, A.; Barigelletti, F.; Balzani, V.; Belser, P.; von Zelewsky, A. 1982. Isr. J. Chem., 22, 87.
Juris, A.; Barigelletti, F.; Campagna, S.; Balzani, V.; Belser, P.; von Zelewsky, A. in press. Coord. Chem. Rev.
Maestri, M.; Sandrini, D.; Balzani, V.; Maeder, U.; von Zelewsky, A. 1987. Inorg. Chem., 26, 1323.
Ohsawa, Y.; Sprouse, S.; King, K.A.; DeArmond, M.K.; Hanck, K.W.; Watts, R.J. 1987. J. Phys. Chem., 91, 1047.
Pina, F.; Ciano, M.; Moggi, L.; Balzani, V. 1985a. Inorg. Chem., 24, 844.
Pina, F.; Mulazzani, Q.G.; Venturi, M.; Ciano, M.; Balzani, V. 1985b. Inorg. Chem., 24, 848.
Reveco, P.; Cherry, W.R.; Medley, J.; Garber, A.; Gale, R.J.; Selbin, J. 1986. Inorg. Chem., 25, 1842.
Rodriguez-Ubis, J.-C.; Alpha, B.; Plancherel, D.; Lehn, J.-M. 1984. Helv. Chim. Acta, 67, 2264.
Watts, R.J. 1983. J. Chem. Educ., 60, 834.

REACTIVITY OF CO_2 AND RELATED HETEROCUMULENES TOWARDS TRANSITION METAL COMPOUNDS

E. Carmona, A. Galindo, M.A. Muñoz

Departamento de Química Inorgánica
Instituto de Ciencias de Materiales
Facultad de Química. Universidad de Sevilla-CSIC.
Aptdo. 553, Sevilla. Spain.

ABSTRACT

The reactivity of carbon dioxide and carbon disulphide towards various molybdenum and tungsten complexes containing dinitrogen, N_2, and ethylene, C_2H_4, ligands, will be described. Action of CO_2 (4-5 atm, 20°C) upon a solution of the complex cis-$Mo(N_2)_2(PMe_3)_4$ produces the bis(carbon dioxide) adduct trans-$Mo(CO_2)_2(PMe_3)_4$, as well as two carbonyl-carbonate complexes, $Mo(CO_3)CO(PMe_3)_4$ and $|Mo(CO_3)CO(PMe_3)_3|_2$, resulting from the reductive disproportionation of two molecules of CO_2 induced by the metal complex: $2CO_2 + 2e = CO_3^{2-} + CO$. The bis(carbon dioxide) adduct has been fully characterized by IR and NMR studies and by conversion into the isocyanide derivatives trans,mer-$Mo(CO_2)_2(CNR)(PMe_3)_3$ (R= Me, i-Pr, t-Bu...), two of which have been structurally authenticated by X-ray crystallography.

The coupling of a coordinated ethylene ligand with a molecule of CO_2 to form acrylic acid has been observed in the reaction of CO_2 with the complexes trans-$M(C_2H_4)(PMe_3)_4$ (M=Mo, W), to yield the new compounds $|M(CH_2=CH-CO_2)H(C_2H_4)(PMe_3)_2|_2$. Hydrogenation of these species provides the corresponding propionate derivatives $|M(O_2CCH_2CH_3)H_n(PMe_3)_4|$ (n=1, M=Mo; n=3, M=W), from which the propionate fragment can be separated, as the lithium salt, by action of Bu^hLi. Finally the interaction of the ethylene complexes $M(C_2H_4)_2(PMe_3)_4$ with CS_2, and some reaction chemistry of the resulting products will be described.

INTRODUCTION

The first step in the activation of carbon dioxide by transition metal compounds is the formation of a $M-CO_2$ complex, since it is through coordination that the electronic structure of this molecule, and hence its reactivity, can be substantially modified. Transition metal complexes containing carbon dioxide in its intact form have received considable attention in the last decade(Inone et al, 1982), mainly with the aim of finding model systems for the activation of CO_2 and subsequent transformation into organic chemicals of comercial interest (Aresta et al, 1987). Despite considerable and intensive work in this area, the number of structurally characterized carbon dioxide complexes is still very limited, and they have been found to contain side-on (Alvarez et al, 1986), η^2-coordinated and η^2, C-coordinated (Calabrese et al, 1983) CO_2. In addition two examples have been reported of what is generally known as assisted coordination of carbon dioxide (Gambarotta et al, 1982), (Herskovitz et al, 1976).

In this paper we present an account of the results obtained by our group in this area. We will mainly emphasize those concerning the coordination of CO_2 in its intact form (Alvarez et al, 1986) as well as those leading to the coupling of this molecule with unsaturated organic substrates, in particular the metal-induced carboxylation and hydrocarboxylation of ethylene to form acrylate and propionate derivatives, respectively (Alvarez et al, 1985).

RESULTS AND DISCUSSION

1. Adduct formation.

Interaction of CO_2 with petroleum ether solutions of cis-$Mo(N_2)_2$ $(PMe_3)_4$ results in the formation of the pale yellow complex trans-$Mo(CO_2)_2$ $(PMe_3)_4$ and of variable amounts of the carbonyl-carbonate derivatives $Mo(CO_3)$ $CO(PMe_3)_4$ and $|Mo(CO_3) CO(PMe_3)_3|_2$. The yellow complex is a microcrystalline, moderately air stable solid, which can be heated at 40-50°C for several hours without noticeable decomposition. Its infrared spectrum displays bands at 1670, 1155 and 1100 cm^{-1}, which by comparison with the spectrum of a sample 50% enriched in $^{13}CO_2$ can be assigned to vibrations associated with the coordinated CO_2 molecules, but no conclusions as to the coordination mode of the CO_2 ligand can be inferred from this data.

^{31}P and ^{13}C NMR data are more informative. Thus the $^{31}P\{^1H\}$ NMR spectrum of the complex is a sharp singlet at 50°C, which converts into a pattern of lines characteristic of an AA'BB' spin system upon cooling at -40°C. On the other hand the ^{31}P NMR of a sample of the complex, approximately 50% enriched in $^{13}CO_2$, clearly shows the three central lines of the 1:4:6:4:1 quintet expected for an isotopic mixture of the three possible isotopomers, while the ^{13}C spectrum of this enriched sample displays a quintet at δ 206 ppm ($^2J_{CP}$=18Hz) due to the CO_2 ligands. A full description of these NMR properties can be found elsewhere (Alvarez et al, 1986). From these data, structure $\underline{A}$ can be proposed for this

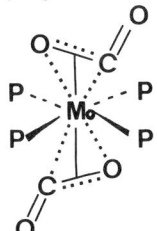

$\underline{A}$

complex, in which the coordinated C=O bonds are staggered with respect to one another and eclipsed with regards to the trans-P-Mo-P vectors of the equatorial plane. A fluxional process, probably involving rotation of the CO_2 ligands is clearly responsible for the temperature dependence of the NMR spectra.

Interaction of the bis (carbon dioxide) adduct with various isocyanides affords new carbon dioxide complexes of composition trans, mer-$Mo(CO_2)_2$(CNR)(PMe$_3$)$_3$, (R=Me, i-Pr, t-Bu, C_6H_{12}, $CH_2C_6H_5$), as shown in eq.1. In no case has substitution of a second PMe$_3$ or of a CO_2 ligands been observed.

$$Mo(CO_2)_2(PMe_3)_4 + CNR \longrightarrow Mo(CO_2)_2(CNR)(PMe_3)_3 + PMe_3 \quad (1)$$

With the exception of some absorptions arising from the coordinated CNR group, the IR spectra of the two complexes are very similar. In particular the bands at 1670, 1155 and 1100 cm^{-1}, associated with the CO_2 ligands, have nearly the same frequencies, and this indicates that the phosphine substitution reaction takes place without change in the coordination mode of the CO_2 ligands. This has been confirmed by NMR studies and by X-ray structural determinations, carried out with the iso-propyl and benzyl derivatives, which have the structure shown schematically in B.

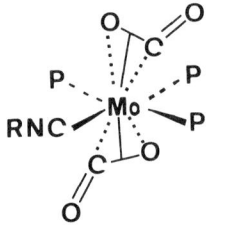

B

2. CO_2 coupling reactions

The use of CO_2 as a one-carbon precursor of organic chemicals is one of the most important goals of modern organometallic chemistry. In the course of our studies on CO_2 chemistry, we have found that the complexes trans-$M(C_2H_4)_2(PMe_3)_4$ (M=Mo,W), react under very mild conditions with CO_2, to produce white, crystalline solids. Infrared and NMR studies suggest that coupling of CO_2 with one of the ethylene molecules has ocurred, with formation of acrylic acid or acrylate ligands, as depicted in eq 2. This process

$$M(C_2H_4)_2(PMe_3)_4 + CO_2 \longrightarrow \tfrac{1}{2}|M(H_2C=CH-CO_2)H(C_2H_4)(PMe_3)_2|_2 \quad (2)$$

corresponds formally to the insertion of CO_2 into one of the ethylene C-H bonds. On the other hand, hydrogenation of the complexes in the presence of PMe_3 produces propionate complexes of composition $M(O_2CCH_2CH_3)H_n(PMe_3)_4$ (n=1, M=Mo; n=3, M=W). The overall process can be considered, in a formal sense, as the hydrocarboxylation of ethylene induced by the metal complex (eq. 3).

$$CH_2=CH_2 + CO_2 + H_2 \longrightarrow CH_3CH_2COOH \quad (3)$$

We haved attempted carrying out similar coupling reactions using other olefine complexes. Reduction of $MCl_4(PMe_3)_3$ (M=Mo,W) with Na-Hg, under propylene, give yellow complexes of composition $M(C_3H_6)(PMe_3)_4$. The appearance of an IR absorption of these compounds at ca. 1700 cm^{-1} and of a hydride resonance in the ^1H NMR spectrum (δ -4.7 quintet, $^2J_{H-P}$ = 50Hz) clearly indicate that these compounds are best formulated as hydrido-allyl derivatives, $(\eta^3-C_3H_5)MH(PMe_3)_4$. Both compounds react wich CO with reductive elimination of propylene and formation of mixed carbonyl-phosphine complexes, $M(CO)_n(PMe_3)_{6-n}$. Rather disappointingly, the reactions of both compounds with CO_2 take place without coupling of this molecule with the unsaturated organic ligand, only reductive elimination of propylene being observed. The inorganic products are $Mo(CO_2)_2(PMe_3)_4$, $Mo(CO_3)CO(PMe_3)_4$ and $|Mo(CO_3)CO(PMe_3)_3|_2$ for M=Mo, and $W(CO_3)CO(PMe_3)_4$ and $WH_2(CO_3)(PMe_3)_4$ for M=W. Doubtless, the small amounts of water present in the CO_2 used, must play an important part in the formation of the hydrido-carbonate complex.

An obvious extension of the above studies concernig the reactivity of olefine complexes towards CO_2 is the investigation of the chemistry of the relate molecule CS_2. Carbon disulphide chemistry is often considered as a model for carbon dioxide-transition metal chemistry. When CS_2 is added to a diethyl ether solution of the ethylene complexes, a black finely divided material is formed which displays a strong IR absorption at 1020 cm^{-1}, due to a CS_2-derived ligand. Analytical and spectroscopic studies indicate the compounds are not related to the CO_2 products but should be formulated instead as $M(C_2S_4)(C_2H_4)(PMe_3)_3$, containing the head-to-tail dimer C_2S_4 group (eq. 4), formed by the reductive dimerization of two molecules of CS_2 induced by the metal complexes. This has been confirmed by an X-ray single crystal determination carried out with the molybdenum complex, for which the structure schematically represented in C has been found. Work is in progress to ascertain the chemical reactivity and other properties of these complexes.

$$\underset{M = Mo, W}{\xrightarrow{2\ CS_2}} \quad (4)$$

C

Acknowledgment.

We gratefully acknowledge generous support of this work by the Comisión Asesora de Investigación Científica y Técnica.

REFERENCES

Inoue, S. and Yamazaki, N. Eds.1982. "Organic and Bioorganic Chemistry of Carbon Dioxide" (Wiley, New York).

Aresta, M. and Forti, G. Eds. 1987. "Carbon Dioxide as a source of Carbon. Biochemical and Chemical Uses". (Reidel Publishing Company, Holland).

Alvarez, R.; Carmona, E. ; Marín, J.M.; Poveda, M.L.; Gutierrez-Puebla, E. and Monge, A. 1986. J. Am. Chem. Soc. , 108, 2286-2294.

Calabrese, J.C. ; Herskovitz, T. and Kinney, J.B. 1983. J. Am. Chem. Soc., 105, 5914.

Gambarotta, S. ; Arena, F. ; Floriani, C. and Zanazzi, P.F. 1982. J. Am. Chem. Soc.,104, 5082.

Herskovitz, T. ; Guggenberger, L. 1976. J. Am. Chem. Soc., 98, 1615.

Alvarez, R.; Carmona, E. ; Cole-Hamilton, D.; Galindo, A.; Gutierrez-Puebla, E.; Monge, A. ; Poveda, M.L. and Ruiz, C. 1985. J. Am. Chem. Soc. 107, 5529-5531.

MEDIUM EFFECTS UPON THE STABILITY OF n-GaAs-BASED
PHOTOELECTROCHEMICAL CELLS

S. Lingier[*], D. Vanmaekelbergh[**] and W.P. Gomes[*]
[*] Rijksuniversiteit Gent, Laboratorium voor Fysische Scheikunde
Krijgslaan 281, B-9000 Gent (Belgium)
[**] Rijksuniversiteit Utrecht, Fysisch Laboratorium
P.O.Box 80000, NL-3508 TA Utrecht (The Netherlands)

ABSTRACT

The competition between the anodic dissolution of an illuminated n-GaAs electrode and the anodic oxidation of tetramethyl-p-phenylenediamine (TMPD) was studied by rotating ring-disk voltammetry as a function of the photocurrent density, the TMPD concentration and the water activity in the electrolyte solution. The water activity was varied by using aqueous solutions of different electrolyte concentration and by using mixed water + methanol and mixed water + acetonitrile solvents of different compositions. The kinetic results can be interpreted on the basis of a chemical step occurring after the first electrochemical step of the decomposition process of GaAs, in which a valence band hole reacts with a surface bond to form a mobile surface intermediate. The chemical step is an equilibrium reaction in which a water molecule reacts with the mobile intermediate leading to the formation of an immobile O-containing surface species and a dissolved proton.
From the results it is concluded that the equilibrium depends on the electrolyte composition and has an influence on the surface concentration of the mobile and immobile decomposition intermediates. Hence, the kinetics of the photooxidation of TMPD and the decomposition at illuminated n-GaAs photoanodes are dependent on the electrolyte composition.

INTRODUCTION

In connection with the potential use for solar energy conversion purposes, several groups have studied the competition at illuminated n-GaAs electrodes, between the anodic oxidation of dissolved reducing agents (this is called the stabilization reaction) and the anodic decomposition of the semiconductor itself. This competition can be quantitatively expressed by the stabilization ratio s, i.e. the fraction of the photocurrent corresponding to the anodic oxidation of a dissolved reducing agent. Specifically in the case of the anodic dissolution of GaAs, which constitutes the reaction of interest in the present paper, Gerischer has shown experimentally that the oxidation of each GaAs unit involves six positive holes and has suggested that water molecules also participate in the reaction in acid medium (Gerischer, 1965). The same author presented a general theoretical discussion on the mechanisms of anodic oxidation reactions of semiconductors, including possible chemical steps (Gerischer, 1970). Kinetic studies on electrode stabilization in aqueous electrolyte solutions (for a review, see Miller, 1984) have thus far contributed to a better insight into the electrochemical mechanisms of stabilization and decomposition, but have hitherto provided little information on the chemical steps occurring in the dissolution process. However, study of the competition between photoanodic decomposition of a GaAs-electrode and the oxidation of tetramethyl-p-phenylevediamine (TMPD) in aqueous solutions of varying pH, lead to the conclusion that a chemical step involving a water molecule takes place after the first electrochemical step of the photodecomposition (Lingier, 1987):

$$X_1 + H_2O \underset{k_{-c}}{\overset{k_c}{\rightleftharpoons}} X_1\text{-OH} + H^+ \qquad (1)$$

where X_1 stands for a positively charged decomposition intermediate which is mobile in a two dimensional surface layer formed by reaction of a valence band hole with a GaAs-surface bond. Reaction of X_1 with a water molecule leads to the formation of an uncharged immobile intermediate X_1-OH and a proton. The second electrochemical step was assumed to be as follows:

$$X_1 + X_1\text{-OH} \overset{k_2}{\rightarrow} X_2 + (\text{GaAs})_{surf} \qquad (2)$$

It was furthermore assumed that oxidation of TMPD occurs by intermediates of the type X_2. With these hypotheses, the strong pH-dependence of the competition between photodecomposition and oxidation of TMPD can be explained quantitatively (Lingier, 1987). As the chemical step (1) involves a water molecule, it may be expected that variation of the water activity will also influence the kinetics of stabilization and decomposition. Variation of the water activity can be achieved by varying the electrolyte concentration over a wide range (Schneemeyer, 1982) and by changing the organic solvent content in water + organic solvent mixtures. In the present paper, results will be presented of stabilization measurements performed in aqueous solutions with high electrolyte concentration, in water + methanol solvents and in water + acetonitrile solvents of various composition. The results of the stabilization experiments performed in water + methanol solvents have already been published (Lingier, 1987), therefore, these results will be reported here very briefly.

EXPERIMENTAL

The measurements were performed on n-type GaAs single crystals purchased from Metallurgie Hoboken-Overpelt (Belgium). According to the specifications of the manufacturer, the density N_D of donors (Sn) ranged between

4.2×10^{16} and 5.5×10^{16} cm^{-3}. The ($\bar{1}\bar{1}\bar{1}$) face of the crystals was exposed to the electrolyte. For details on the preparation of the ohmic contact and on the mounting of the sample into a rotating ring-disk (RRD) set-up, see Vanmaekelbergh, 1985. The counter electrode of the cell circuit was a Pt-gauze and the reference electrode was an aqueous saturated sulphate electrode (SSE).

The indifferent electrolyte solution consisted of either water, a water + methanol mixture or a water + acetonitrile mixture in which 0.25 mol.dm^{-3} LiClO$_4$ was dissolved. For the experiments in aqueous medium under high electrolyte concentration LiCl was prefered because high LiClO$_4$ concentration favours the formation of a passivating oxide layer on the GaAs-surface.

The pH of the solutions was determined by a glass electrode, which was calibrated by means of standard buffer solutions. It should be recalled that, in principle, proton activities determined in this way in solutions with high electrolyte concentrations and different H$_2$O/CH$_3$CN and H$_2$O/CH$_3$OH ratios cannot be mutually compared.

The stabilizing agent used was TMPD. Dissolved TMPD may react with either one or two protons to form TMPDH$^+$ and TMPDH$_2^{2+}$, respectively. In the following, TMPDH$_2^{2+}$, TMPDH$^+$ and TMPD will be denoted by Y_0, Y_1 and Y_2 respectively. Only Y_1 ions and Y_2 molecules are expected to act as reducing agents (Nickel, 1983 and Nickel, 1985). The Y_0, Y_1 and Y_2 equilibrium concentrations (y_0, y_1 and y_2) were calculated from the concentration c of dissolved TMPD and from the acidity constants K_1 and K_2 of Y_0 and Y_1, which were determined for each solvent from titrations of TMPD with HClO$_4$.

The stabilization ratio s was determined by rotating ring-disk voltammetry; the product formed from TMPD (or TMPDH$^+$) oxidation at the GaAs disk under illumination was determined quantitatively by reduction at the gold ring. Before the experiments, the GaAs electrode was etched chemically and photoelectrochemically (Vanmaekelbergh, 1985). Illumination of the electrode was provided by a halogen incandescent lamp. The light intensity was varied by means of neutral density filters. All measurements were performed at room temperature (298 K).

RESULTS

General

Measurements of the stabilization ratio s were performed on the GaAs photoanode in aqueous medium with 0.25 mol.dm^{-3} and with 4 mol.dm^{-3} LiCl, in three water + methanol mixtures with 18, 48 and 80 mol % methanol respectively, and in two water + acetonitrile mixtures with 13 and 42 mol % acetonitrile. The stabilization ratio s was measured as a function of the photocurrent density i and the concentration c of dissolved TMPD. The measurements were performed at a constant electrode potential V corresponding to high band bending, so that surface recombination can be neglected. All experiments were performed in acid medium as required for the solubility of TMPD and decomposition products of GaAs.

Aqueous solution with 0.25 mol.dm^{-3} LiClO$_4$

The relationship between s, i and c corresponds to the following kinetic law:
$$s^2/6(1-s) + s^3/(1-s)^2 = l.(c^2/i) \qquad (3)$$
with l a proportionality factor.
Eqn. (3) corresponds to one of the reaction schemes formerly proposed in the framework of a general kinetic analysis (Cardon, 1980).

Aqueous solution with 4 mol.dm^{-3} LiCl

The relationship between s, i and c was measured at pH 2.0 and 3.0. It appears that s increases with increasing c, but decreases with increasing i and hence increasing light intensity. The following kinetic law, observed previously for the system n-GaAs/Fe^{2+}(aq) (Vanmaekelbergh, 1985), was found to hold:
$$s^2/(1-s) = k.(c^2/i) \qquad (4)$$
In fig. 1 the data have been replotted as log [$s^2/(1-s)$] vs. log (c^2/i), yielding a straight line of unit slope. The values of the proportionality factor k were 1.3 x 10^3 and 7.1 x 10^4 C.cm^4.s^{-1}.mol^{-2} at pH 2.0 and 3.0 respectively.

Mixed solvent with 18 mol % CH$_3$OH

In this medium, the relationship between s, i and c is the same as found in aqueous medium of low electrolyte concentration.

Mixed solvents with 48 and 80 mol % CH$_3$OH

Kinetic measurements performed in these two solutions yielded a relationship between s, i and c that obeys the following law:
$$s/6 + s^2/(1-s) = k'.(c/i) \qquad (5)$$

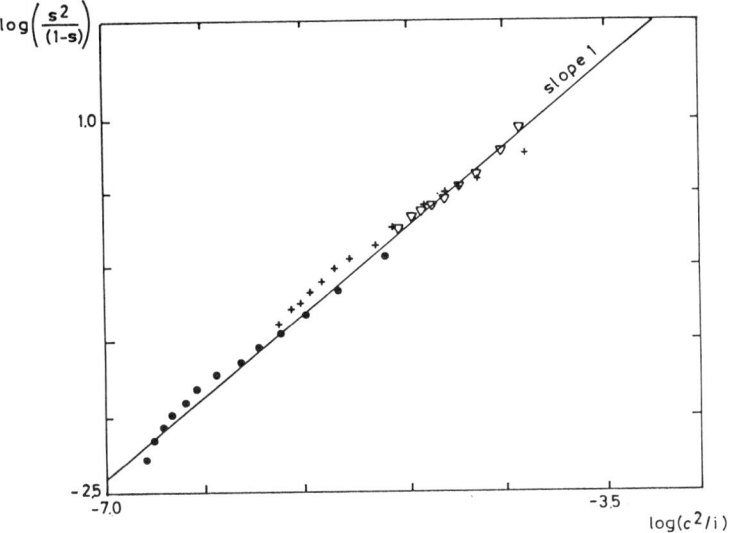

Fig. 1 Log $[s^2/(1-s)]$ vs. log (c^2/i). V = -0.3 V (vs. SSE).
Electrolyte composition: 4 mol.dm^{-3} LiCl + c mol.dm^{-3} TMPD in H$_2$O, pH = 3.0. c : 0.01 (●) , 0.03 (+) , 0.08 ($\triangledown$).

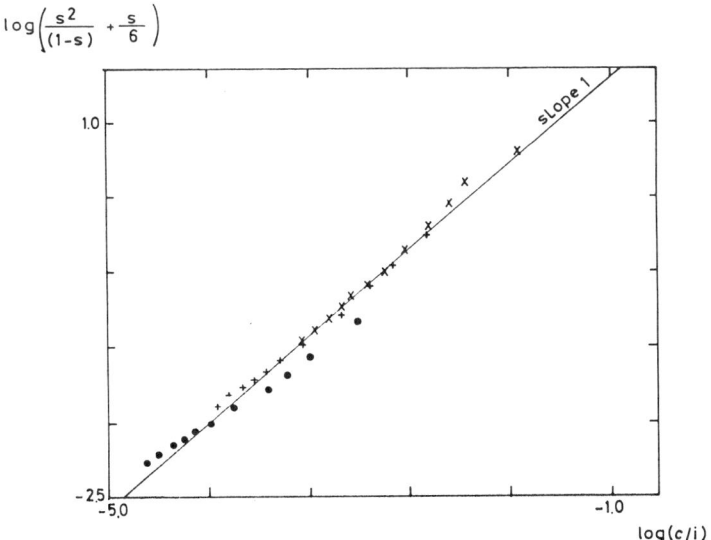

Fig. 2 Log $[s^2/(1-s) + s/6]$ vs. log (c/i). V = -0.3 V (vs. SSE).
Electrolyte composition: 0.25 mol.dm^{-3} LiClO$_4$ + c mol.dm^{-3} TMPD in a H$_2$O + CH$_3$CN mixture with 13 mol % CH$_3$CN, pH = 2.0.
c : 0.11 (●) , 0.041 (+) , 0.202 (x).

Mixed solvents with 13 mol % CH_3CN

In this medium, the relationship between s, i and c obeys eqn. (5) as can be seen in fig. 2. The values of the proportionality factor k' were 6.76×10^4, 1.48×10^5, 2.14×10^6 and 4.37×10^6 $dm.C.mol^{-1}.s^{-1}$ at pH 1.0, 1.5, 2.0 and 2.5 respectively.

Mixed solvents with 42 mol % CH_3CN

The kinetics measured in this medium obey eqn. (4). This can be seen in fig. 3. The values of the proportionaly factor k are 25.1, 218.8 and 309 $C.cm^4.mol^{-2}.s^{-1}$ at pH 1.5, 2.0 and 2.5 respectively.

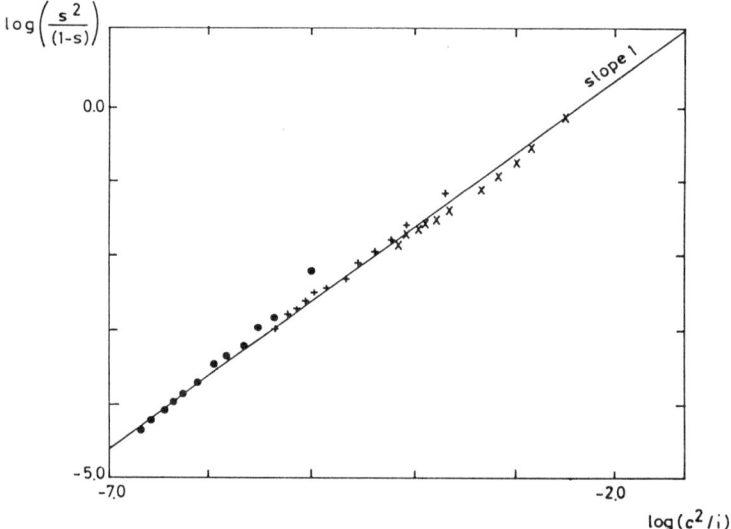

Fig. 3   Log $[s^2/(1-s)]$ vs. log (c^2/i). V = -0.3 V (vs. SSE). Electrolyte composition: 0.25 $mol.dm^{-3}$ $LiClO_4$ + c $mol.dm^{-3}$ TMPD in a H_2O + CH_3CN mixture with 42 mol % CH_3CN, pH = 2.0.
c : 0.011 (•) , 0.050 (+) , 0.205 (x).

DISCUSSION

The kinetic laws (eqn. (3), (4) and (5)), which were found to describe the competition between the photoanodic oxidation of TMPD and that of GaAs, can be interpreted on the basis of a reaction mechanism comprising the following electrochemical steps. Capture of a valence band hole h^+ in a surface bond of the semiconductor $(GaAs)_{surf}$ leads to the formation of an intermediate X_1. In all the following decomposition steps a mobile species (M) participates, which can be an intermediate X_1 or a valence band hole. In this way intermediates of a higher degree of oxidation X_i (i = 2,...,5) and products are formed which go into solution:

$$(GaAs)_{surf} + h^+ \xrightarrow{k_1} X_1 \tag{6}$$

$$X_1 + M \xrightarrow{k_2} X_2 \tag{7}$$

$$X_5 + M \xrightarrow{k_6} \text{decomposition products} \tag{8}$$

As was already mentioned in the introduction, the kinetic law (eqn. (3)), which describes the competition between the oxidation of TMPD and that of GaAs in aqueous solutions of low electrolyte concentration and in water + methanol mixtures with 18 mol % CH_3OH, can be interpreted assuming that the intermediate X_1 is mobile and participates in all electrochemical steps. Moreover, from the pH-dependence we were able to state that reaction (1) occurs forming an immobile intermediate X_1-OH. The following electrochemical step is given by reaction (2). Stabilization, i.e. electrochemical oxidation of the dissolved reducing agent Y_j (j = 1, 2), is assumed to occur by electron transfer from Y_j to the intermediate X_2:

$$X_2 + Y_j \xrightarrow{k^s_{-2}} X_1 + \text{oxidation product} \tag{9}$$

The kinetic law (eqn. (4)), which was found to describe the competition between the photoanodic oxidation of TMPD and that of GaAs in aqueous medium with 4 mol.dm^{-3} and in water + acetonitrile mixtures with 42 mol % acetonitrile, can be interpreted on the basis of a reaction mechanism in which the intermediate X_1 acts as a mobile reactive species in the decomposition steps. Stabilization is assumed to occur by electron transfer from Y_j to the intermediate X_1:

$$X_1 + Y_j \xrightarrow{k^s_{-1}} (GaAs)_{surf} + \text{oxidation product} \tag{10}$$

The net steady-state formation rate of X_2 from X_1 is given by the following expression:

$$(1-s).i/6.e = k_2.x_1^2 \tag{11}$$

and the hole flux associated with the oxidation of Y_j is then:

$$s.i/6.e = k^s_{-1}.y_j.x_1 \tag{12}$$

where x_1 is the surface concentration of X_1 and e is the elementary charge. Eliminating x_1 from eqn. (11) and (12), one obtains:

$$s^2/(1-s) = [(e.k^s_{-1})^2/6.k_2].(y_j^2/i) \tag{13}$$

The relationship between the concentration of one of the forms of TMPD and the analytical TMPD concentration c is given by

$$y_j = c/f_j(K_1, K_2, C_{H+}) \quad (14)$$

in which $f_j(K_1, K_2, C_{H+})$ is a specific function for each form of dissolved TMPD. At a given pH, y_j is proportional to c, and hence eqn. (13) is in agreement with the experimental results. The experimental constant k is then interpreted as being

$$k = e \cdot (k_{-1}^s)^2 / 6 \cdot k_2 \quad (15)$$

In order to investigate whether our model can also explain the observed pH-dependence of k, the value of k (measured in aqueous medium of 4 mol.dm^{-3} LiCl) multiplied by $[f_j(K_1, K_2, C_{H+})]^2$ was listed as a function of pH in table 1.

TABLE 1   Values of $k \cdot [f_j(K_1, K_2, C_{H+})]^2$.

$k \cdot [f_j(K_1, K_2, C_{H+})]^2 / \dfrac{C \cdot cm^4}{s \cdot mol^2}$ Y$_j$	pH = 2.0	pH = 3.0
j = 1 (TMPDH$^+$)	7.8 × 10^5	7.9 × 10^5
j = 2 (TMPD)	8.6 × 10^{18}	9.5 × 10^{12}

In the case Y_j = TMPDH$^+$, $k \cdot [f_j(K_1, K_2, C_{H+})]^2$ was independent of the pH. The same conclusion can be made for the water + acetonitrile mixture with 42 mol % CH$_3$CN. From the foregoing discussion, it is clear that the kinetic law (4), observed in aqueous solution with 4 mol.dm^{-3} LiCl and also in water + acetonitrile solutions with 42 mol % CH$_3$CN, can be explained by a model involving the subsequent decomposition steps (6), (7), (8) and the stabilization reaction (10) in which Y_j stands for TMPDH$^+$.

The kinetic law (eqn. (5)), which was found to describe the competition between the photoanodic oxidation of TMPD and that of GaAs in water + methanol mixtures with 48 and 80 mol % CH$_3$OH and in water + acetonitrile mixture with 13 mol % CH$_3$CN, can be interpreted on the basis of a reaction mechanism in which the subsequent electrochemical steps in the anodic decomposition of the semiconductor occur by the capture of a free hole at each step (instead of by reaction with mobile intermediates X$_1$, as was the case with the mechanisms discussed before). The oxidation of the dissolved reducing agent occurs through reaction with X_1 or X_1-OH.

The change in decomposition mechanism, i.e. the nature of the mobile species, on passing from one solvent to another solvent can be interpreted on the basis of reaction (1) which we proposed to explain the pH-dependence of the results in aqueous medium at low electrolyte concentration (Lingier, 1987).

In both aqueous solutions of high and low electrolyte concentrations and in water + methanol mixtures with 18 mol % CH_3OH the intermediate X_1 acts as a mobile reactive species.

Because of the drastically reduced water activity in aqueous medium of 4 mol.dm^{-3} LiCl, equilibrium (1) is lying to the left, in favour of the mobile intermediates X_1. The concentration of X_1-OH is then decreased drastically. This can be the cause why the second decomposition step consists of the reaction between two positively charged intermediates X_1 (instead of the reaction between a charged mobile intermediate X_1 and an uncharged immobile intermediate X_1-OH found in aqueous medium of low electrolyte concentration).

At present we are unable to explain the change in stabilization mechanism on passing from aqueous medium of low electrolyte concentration and water + methanol mixtures of 18 mol % CH_3OH (reaction with X_2) to the other solvents (reaction with X_1). Possibly, the chemical nature and reactivity of X_2 formed by reaction (2) in aqueous medium of low electrolyte concentration and in water + methanol mixtures with 18 mol % CH_3OH differs from the chemical nature of X_2 formed by the reaction (7) in aqueous medium under high electrolyte concentration.

The fact that valence band holes take over the role of the mobile reactive species from the intermediates X_1 in water + methanol mixtures with 48 and 80 mol % CH_3OH and in water + acetonitrile mixtures with 13 mol % CH_3CN can be attributed to a shift in equilibrium (1) due to a medium effect, leading to the immobilization of a major part of the intermediates X_1. Indeed, according to literature data (Feakins, 1963; Case, 1967; Das, 1981), the standard Gibbs energy of the transfer of the proton from water to water + methanol and water + acetonitrile mixtures, $\Delta G^o_{t,H^+}$, as a function of the methanol and acetonitrile content, exhibits a minimum around 80 mol % CH_3OH and 22 mol % CH_3CN. Hence, when the methanol or acetonitrile content increases to these values, $\Delta G^o_{t,H^+}$ provides an increasingly negative contribution to the ΔG^o value of reaction (1), which has the effect of increasing the corresponding equilibrium constant and hence of shifting equilibrium (1) to the right lowering x_1 in favour of the concentration of immobile intermediates. Because of the small contribution of $\Delta G^o_{t,H_2O}$ (the standard Gibbs energy of transfer of the water molecule from water to water + methanol or water + acetonitrile mixtures) with respect to that of $\Delta G^o_{t,H+}$, $\Delta G^o_{t,H_2O}$ has not to be taken into account. Nevertheless, in view of the negative values generally cited for the $\Delta G^o_{t,H+}$ in the concentration range in which our experiments were performed, it is reasonable to assume that the medium effect on the proton is able, despite the reduced water activity, to cause a shift of equilibrium (1) to the right which is sufficiently large to explain the change in the decomposition mechanism.

In water + acetonitrile mixtures it seems that the values of $\Delta G^o_{t,H+}$ become positive from 70 mol % CH_3CN and higher (Das, 1981). Taking into account the reduced water activity and the insufficient negative value of $\Delta G^o_{t,H+}$ at acetonitrile content smaller than 70 mol %, we can assume that the equilibrium (1) is not shifted anymore to the right in a water + acetonitrile mixture with 42 mol % CH_3CN. This results in a participation of intermediates X_1 as mobile reactive species in the decomposition steps, instead of valence band holes, in water + acetonitrile mixtures with 42 mol % CH_3CN.

CONCLUSIONS

By varying the water activity in electrolyte solutions via two different ways further evidence was found confirming the existence of a chemical step in the decomposition of n-GaAs electrodes under illumination. In this chemical reaction a surface intermediate, formed by reaction between a valence band hole and a GaAs surface bond, reacts with a water molecule to form an immobile oxygen-containing surface species and a dissolved proton. Techniques other than kinetic ones will be needed to obtain further information on the nature of all other oxygen-containing surface intermediates.

ACKNOWLEDGEMENTS

The support of the I.I.K.W. (Interuniversitair Instituut voor Kernwetenschappen) to this work is gratefully acknowledged. The authors, S.L. and D.V. wish to thank the I.W.O.N.L. (Instituut tot Aanmoediging van het Wetenschappelijk Onderzoek in Nijverheid en Landbouw) and the N.F.W.O. (Nationaal Fonds voor Wetenschappelijk Onderzoek) respectively for a fellowship.

REFERENCES

Cardon, F., Gomes, W.P., Vanden Kerchove, F., Vanmaekelbergh, D. and Van Overmeire, F. 1980. On the kinetics of semiconductor electrode stabilization. Faraday Disc., 70, 153-164.

Case, B. and Parsons, R. 1967. The real free energies of solvatation of ions in some non-aqueous and mixed solvents. Trans. Faraday Soc., 63, 1224-1239.

Das, K., Das, A.K. and Kundu, K.K. 1981. Ion-solvent interactions in acetonitrile + water mixtures. Electrochim. Acta, 26, 471-478.

Feakins, D. and Watson, P. 1963. Studies of ion solvatation in non-aqueous solvents and their aqueous mixtures. Part II. Properties of ion constituents. J. Chem. Soc., 4734-4741.

Gerischer, H. 1965. Uber der Mechanismus der anodischen Auflösung von GaAs. Ber. Bunsenges. Phys. Chem., 69, 578-583.

Gerischer, H. 1970. Physical Chemistry, An Advanced Treatise, vol. 9A (Academic Press, New York), p. 463.

Lingier, S., Vanmaekelbergh, D. and Gomes, W.P. 1987. On the role of chemical steps in the anodic dissolution of illuminated n-type GaAs electrodes. J. Electroanal. Chem., 228, 77-88.

Miller, B. 1984. Charge transfer and corrosion processes at III-V semiconductor/electrolyte interfaces. J. Electroanal. Chem., 168, 91-100.

Nickel, U. and Jaenicke, W. 1982. 1- und 2-Elektronenschritte bei der Oxidation substituierter Paraphenylendiamine mit verschiedenen Oxidationsmitteln. II. Reaktionen mit Hexacyanoferrat. Ber. Bunsenges. Phys. Chem., 86, 695-701.

Nickel, U., Mao Zhou, B. and Gulden, G. 1985. Coupled catalytic and autocatalytic oxidation reactions linked by a feedback reaction. Ber. Bunsenges. Phys. Chem., 89, 999-1004.

Schneemeyer, L.F. and Miller, B. 1982. InP and CdS photoanodes in concentrated aqueous iodide electrolytes. J. Electrochem. Soc., 129, 1977-1981.

Vanmaekelbergh, D., Gomes, W.P. and Cardon, F. 1985. Studies on the n-GaAs photoanode in aqueous electrolytes. 1. Behaviour of the photocurrent in the presence of a stabilizing agent. Ber. Bunsenges. Phys. Chem., 89, 987-994.

UNPINNING OF ENERGY BANDS IN PHOTOELECTROCHEMICAL CELLS:
A CONSEQUENCE OF SURFACE CHEMISTRY AND SURFACE CHARGE

D. Meissner and R. Memming
Institut für Physikalische Chemie der Universität Hamburg
Bundesstraße 45, D-2000 Hamburg 13. FRG

ABSTRACT

A change of the surface chemistry or of the surface charge of an electrode leads to a change of the potential drop across the Helmholtz double layer and by this to unpinning of the band edges at the surface. The result is a different electrochemical and photoelectrochemical behavior. Three cases are considered, which can be distinguished by electrochemical methods: surface charging by minority carrier trapping in the surface, formation of a conducting layer on the surface forming a Schottky barrier or an Ohmic contact, and formation of an insulating layer. However, only surface charging may be advantageous to photoelectrochemistry. A conducting layer leads to a new solid state device with no need to use it in an electrolyte, and an insulating layer only inhibits charge transfer. This is discussed for different materials in order to give criteria for discerning these effects.

INTRODUCTION

Light induced reactions at semiconductor electrodes can be very efficient because electron-hole pairs are easily separated by the electric field across the space charge region below the semiconductor surface. Accordingly, one has a powerful driving force for minority carrier reactions. Theoretically, the photocurrents across the interface are expected to start around the flatband potential U_{fb} at which the field is zero. Experiments with many semiconductors have shown, however, that the onset of photocurrent occurs at potentials differing considerably from U_{fb} in the dark. As we have shown already several years ago, this effect is related to bandedge movement upon illumination (Memming and Kelly, 1981; Kelly and Memming; 1982).

Frequently this leads to severe problems, because it makes it difficult to predict whether a certain semiconductor is suitable to be applied in interesting photoelectrochemical reactions. The consequences of this effect are still not considered sufficiently. Meanwhile this effect was investigated also by other groups (compare f.e. Li and Peter, 1985; Vanmaekelberg, Gomes and Cardon, 1985; Allongue and Cachet, 1985; Kühne and Tributsch 1986) and has proven to be common for various kinds of semiconductors. In the present paper we want to analyse the origin of this effect in more detail.

RESULTS AND DISCUSSION

The position of the bandedges at semiconductor surface in an electrolyte is determined by the potential drop across the interface, i.e. the very first layer of the solid and (in an electrolyte of high conductivity) the Helmhotz double layer in the solution. It changes if charges are accumulated in the surface, or if a chemical reaction changes the surface composition. The question arises, how to investigate the origin of this effect and what are the consequences for photoelectrochemical reactions. According to our present understanding one can distinguish three causes: mainly surface charging (a), formation of a conducting (b) and an insulating layer (c).

(a) Surface State Charging

In I/E curves the onset of photocurrent is expected from classical theories to occur near the flatband potential as measured in the dark ($E_{fb}(d)$), i.e. where the majority carrier current starts too. However, a large shift of the onset potential is seen especially if no additional redox couple is present in the aqueous electrolyte, in cathodic direction for p-, in anodic direction for n-type materials (Fig. 1). This shift depends on the light intensity but saturates already at relatively low intensities (Memming, 1987). If minority carrier acceptors (oxidants for p- and reductants for n-type semiconductors) are added to the solution, the onset can be shifted back to $E_{fb}(d)$ if they have the appropiate redox potential. In principal two types of redox couples can be found: those which lead to a shift of the photocurrent onset potential and those which don't. The transition between the two classes occurs at a specific redox potential.

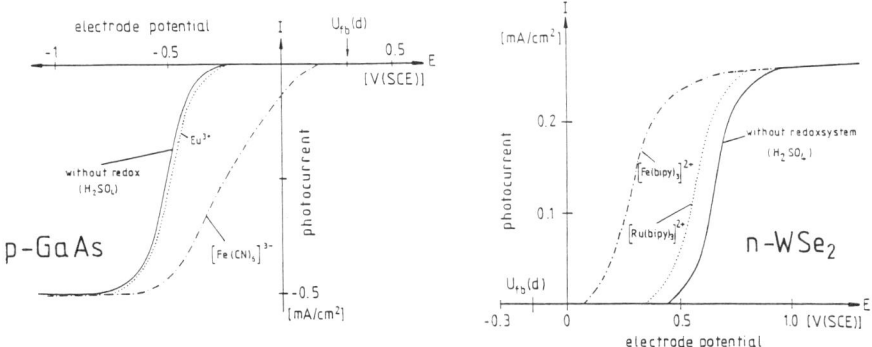

Fig. 1: Photocurrent/potential behavior of p-GaAs and n-WSe$_2$ in aqueous electrolytes with different redox couples

The effect of illumination seen in the current/potential behavior is reflected also in capacity measurements as evaluated in the form of Mott/Schottky-plots (Fig. 2). Illumination leads to a parallel shift of this plot in the same direction and by about the same amount as in I/E curves. The plot is shifted back to its dark position if the appropriate redox couple is added. Other minority carrier acceptors on the other hand are not able to shift the light-plot back onto the plot obtained in the dark.

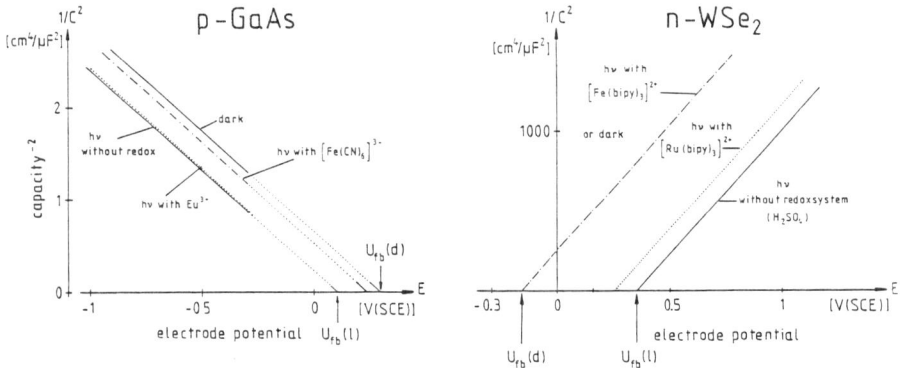

Fig. 2: Mott/Schottky plots of capacity data obtained in the dark and under illumination in different electrolytes

These effects are understood as minority carrier trapping at the surface, which is prevented if suitable minority carrier acceptors are added. In the case of p-GaAs the shifted flatband potential as obtained during illumination, is moved back to its dark value if a redox system such as $[Fe(CN)_6]^{3-}$ or Fe^{2+} is added to the electrolyte (Fig.2). Then the photocurrent occurs already at less cathodic polarization (Fig.1). Since the energy states of an electron acceptor such as $[Fe(CN)_6]^{3-}$ (E_o = +0.1 V(SCE)) are located at rather low energies, the electron transfer is assumed to occur via surface states as indicated in Fig.3. In the absence of this acceptor, the electrons created by light are also captured by surface states. Because an electron transfer is not possible, surface charging occurs leading to an upward shift of the bands.

This model is supported by the result, that the addition of Eu^{3+} does not lead to a backward shift of the flatband potential under illumination although the standard potential of this system ($E_o(Eu^{2+}/Eu^{3+})$ = -0.6 V(SCE)) occurs close to the conduction band. Accordingly, the rate of electron trapping by surface states is much higher than for electron transfer to Eu^{2+}. Similar arguments hold for n-WSe$_2$. Here also the $E_{fb}(l)$ during illumination and the onset potential of photocurrent are identical, and shifted if no redox couple is present. A detailed analysis has shown that in the whole range between U_{fb} (d) and onset potential of photocurrent, the energy bands remain flat and the minority carriers (holes) recombine with electrons.

Corresponding to the anodic shift of the flatband potential upon illumination, the energy bands move downward by about 0.6 eV as illustrated in Fig.3b. Various hole acceptors such as $[Fe(CN)_6]^{4-}$ ($E^o = +0.2$ V(SCE)), Fe^{2+} (+0.4 V) and $Fe(bipy)_3^{2+}$ or $Fe(phen)_3^{2+}$ (both $E^o = +0.9$ V) are capable of moving E_{fb} back to its dark value. The last two couples are of special interest, because their standard potential occurs very close to the valence band (Fig.3b). In this case the hole transfer must occur via states in or close to the valence band. This result is of great importance because a regenerative photovoltaic cell of a high photovoltage can be produced with the system $WSe_2/[Fe(bipy)_3]^{2+/3+}$. Surprisingly, downward movement of bands cannot be avoided by using $[Ru(bipy)_3]^{2+}$ although its standard potential is only slightly lower in energy ($E^o = +1.02$ V) than for $[Fe(bipy)_3]^{2+}$.

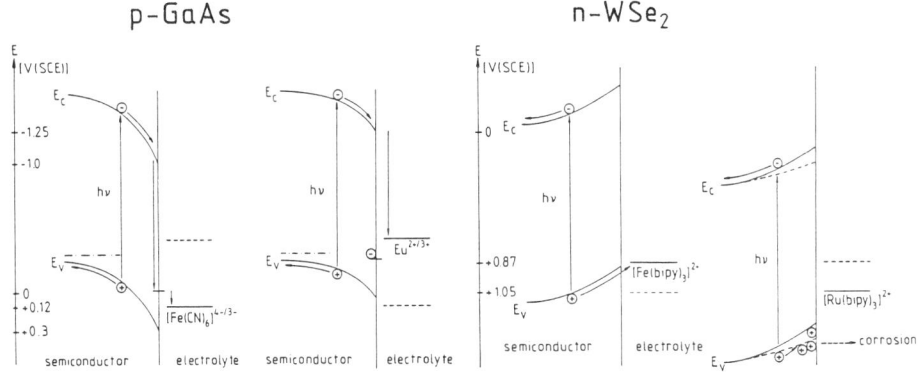

Fig. 3: Energetic model of band-edge shift under illumination for p-GaAs and n-WSe$_2$ with minority carrier acceptors

It should be mentioned further, that the shift of energy bands upon creation of minority carriers does not only occur upon light excitation. Holes can also be produced in the dark via hole transfer from a hole donor into valence band. This kind of process occurs for instance by using the oxidized species of the redox couple $[Ru(bipy)_3]^{2+/3+}$. A corresponding cathodic dark current starts at the same potential at which the photocurrent onset was found in the presence of the reduced form of the same couple. This shift of E_{fb} by hole injection from the electrolyte has been found also with Ce^{4+} at n-WSe$_2$ (McEvoy et al., 1985) and n-GaAs (Schröder et al., 1985).

(b) Formation of a Conducting Layer

As already mentioned above, Eu^{3+} is not capable of preventing surface state charging by capturing electrons excited into the conduction band of p-GaAs although its energy states occur mainly at the conduction band. On the other hand the oxidation of Eu^{2+} is a very efficient process: the anodic (diffusion controlled) dark current starts already at a very cathodic potential (Fig. 4a). This process was proven to be a valence band process (Meissner et al., 1986). The fact that reduction and oxidation can proceed via different bands is due to the wide distribution of energy states (high reorientation energy) in this redox system (Fig.5).

The situation is quite different for InP, the energy bands of which are located at about the same position as those of GaAs. In this case, however, the electrons excited in p-InP are easily transferred to Eu^{3+} and no shift of bands occurs. The oxidation of Eu^{2+}, however, does not take place at p-InP: no anodic dark current was seen (Tubbesing et al., 1986). This result was interpreted by the existence of an In_2O_3-layer. This oxide layer is difficult to remove (Heller et al. 1983). It is a large bandgaps semiconductor. Provided that the conduction band of In_2O_3 is located below that of InP, then electrons can efficiently be transferred to Eu^{3+} as shown in Fig.6. Any

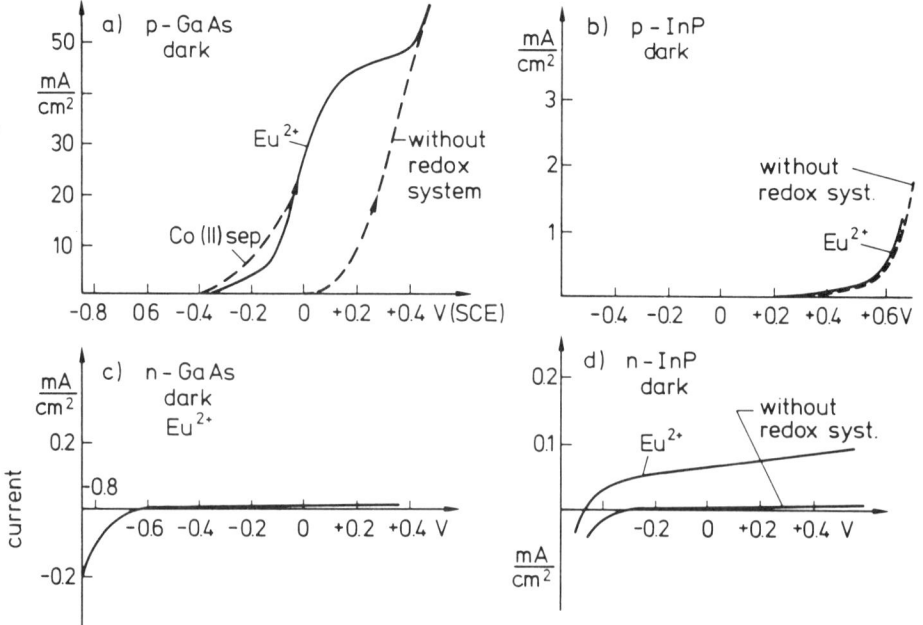

Fig. 4: Dark current/potential behavior of GaAs and InP (both n- and p-type) in the presence of Eu^{2+}

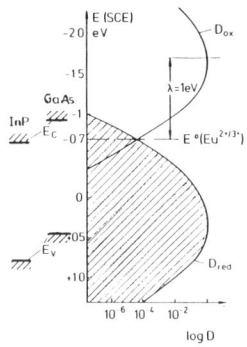
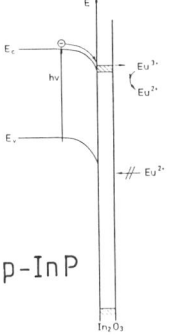

Fig. 5: Distribution of energy states of Eu^{2+}/Eu^{3+} the interface to InP and to GaAs.

Fig. 6: Energy levels of InP and a thin layer of In_2O_3 at the contact to a solution containing $Eu^{2+/3+}$.

transfer via the valence band of InP is not possible because no states exist in the oxide at this energy. Here surface chemistry plays an important role.

Another example for the same kind of effect is $n-RuS_2$ studied extensively by Kühne and Tributsch (1986). For this material (bandgap 1.3 eV) the valence band is located in the dark at about 0 eV (SCE). The bands are shifted downwards by as much as 2 eV upon illumination. Interestingly, the

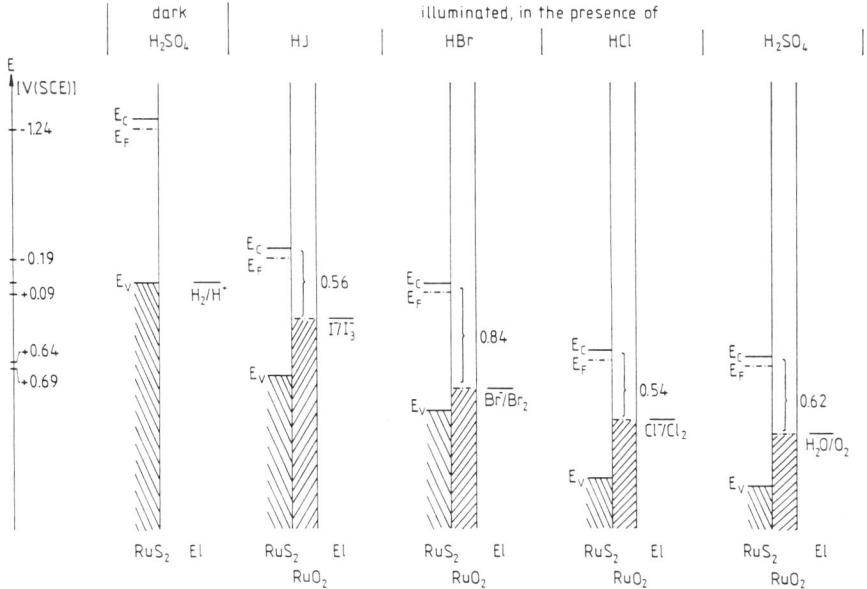

Fig. 7: Flatband situation for RuS_2 in different electrolytes as measured by Kühne and Tributsch (1985), interpreted by us as due to a RuS_2/RuO_2-Schottky junction pinned to the electrolyte redox potential

photocurrent resulting in O_2 evolution occured around +1.1 V(SCE), the valence band being located, however, at +2.3 eV, i.e. considerably below the E^o(H_2O/O_2)-value. The bandedge movement is seen best in capacity measurements which give flatband potentials as shown in Fig. 7.

Adding halides to the electrolyte, the shift of bands under illumination and thereby the photocurrent onset potential decreased in accordance with the change in the redoxpotential of the halide couple used. The results can only be understood in terms of Fermi level pinning by the redox couple about 0.5 to 0.8 V below the Fermi level of the semiconductor. This is expected for a Schottky barrier to a metallic surface layer (RuO_2) coupling to the redox potential of the electrolyte (Fig. 7), as also discussed by Kühne (1987). This means, however, that hole transfer is always connected with an energy loss of more than 0.5 eV. In addition the question arises, whether the semiconductor/metal Schottky junction which acts as a pure solid state device, should really be immersed directly into the electrolyte.

(c) Formation of an Insulating Layer

Photocorrosion frequently leads to nonconducting products which cover the surface. This is known best for silicon, but is also true for GaAs and for the chalcogenides. In some cases the layer can be removed continously by appropiate choise of the solvent, e.g. by using HF for Si, strong acidic or alcaline solution for GaAs or using polychalcogenide solution for CdS and CdSe. For the latter case these solutions have the advantage, that the semiconducting material is even restored, because the corrosion product Cd^{2+} reacts immediately after formation with the chalcogenide to give back the semiconducting material.

We investigated intensively the surface chemistry of CdS. Here the influence of sulfur formed on the surface either by the etching process used to clean the surface from metal deposits, or by the photocorrosion process, was demonstrated by photoelectrochemical as well as by surface spectroscopic methods (Meissner et al., 1978a-c). Even partial sulfur coverage of the surface leads to a dramatic shift of the apparent flatband potential as found by superficially evaluating capacity measurements. However, more careful investigations prove, that the Mott/Schottky plots of these measurements don't really give a straight line (Fig. 8). Also a strong frequency dispersion of these measurements is seen. Both results proof that the equivalent electric circuit, used to obtain the capacity values from the ac-current measurents, is not valid, i.e. that a Mott/Schottky plot may not be used for evaluating the band position. As a consequence it may not be concluded from measurements as shown in Fig. 8a, that the bands at a sulfur covered surface are located more favorable for oxidation processes than on a clean surface (Fig. 8b). Here, for a sulfur free surface, one gets straight plots over a wide potential range, as well as a perfect diode behavior in current/potential measurements. However, the band position is much higher than assumed previously. As a consequence this material is not able to oxidize water.

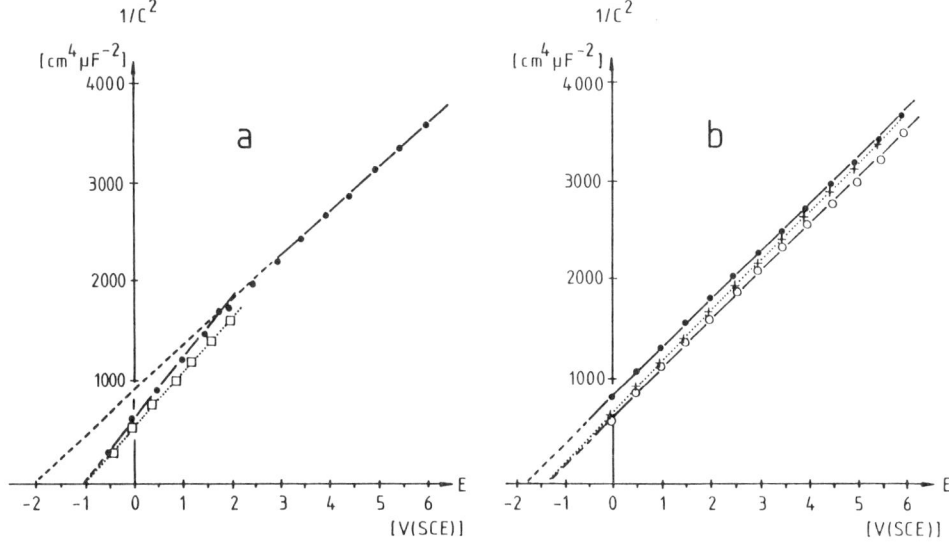

Fig. 8: Mott/Schottky plots as obtained from CdS-(0001)Cd
(a) polished and etched as usual (sulfur contaminated)
(b) sulfur free (for details see Meissner et al. 1987b)

INTERPRETATIONS AND MODELS

As discussed above only surface charging is of real importance for the interpretation of charge transfer at the semiconductor/electrolyte interface. The formation of a conducting layer leads to a solid state device, for which there is no need to be placed directly into the electrolyte. An insulating layer, on the other hand, can not improve charge transfer, except perhaps for very thin layers that allow electron tunneling. The band position at oxidized parts of the surface is not known.

Surface charging has been investigated in more detail using CdS single crystal electrodes (Meissner et al. 1987c), for which a microscopic model was developed based on minority as well as majority carrier charge transfer via surface states. This model can't be discussed here in detail. However, from energy models drawn in analogy to Fig. 3 for this material (Fig. 9) it may be concluded on a first sight, that at a positively charged surface, hole transfer from the valence band may be favorable compared to an uncharged surface. Especially for CdS this is of critical importance, because here the question arises, whether oxygen may in principal be produced. However, from all we know in the moment we would clearly deny this.

Capacity measurements of clean CdS electrodes proof that reversible surface charging is easy to distinguish from irreversible surface oxidation. The latter process is seen to be subsequent to the reversible step. From detailed investigations of corrosion processes of CdS, we attribute the charge trapping to the formation of $S^{-\cdot}{}_s$ from a $S^{2-}{}_s$ surface state. Surface oxidation then follows by a second hole leading to S^o. From $S^{-\cdot}{}_s$ the hole can be further transfered to a reduced species in the solution, which then prevents corrosion. However, no species having a redox potential below the position of the positively charged surface state $(S^{-\cdot}{}_s)$ is capable of preventing photocorrosion. This has been proven experimentally (Memming, 1978; compare also Inoue et al., 1977).

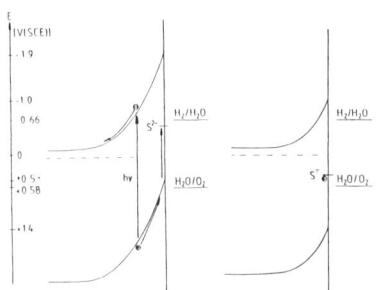

Fig. 9: Energy model for n-CdS based on measurements of the clean surface: Band edge shift by $S^{-\cdot}$ formation relative to the redox potentials of water

This behavior is in very good agreement with our understanding of the charged surface state as beeing $S^{-\cdot}{}_s$. Once a hole is localized at the surface (trapped by surface sulfide $S^{2-}{}_s$), a second hole reaching the same spot in the surface breaks the last bond of the sulfur radical, forming the corrosion product $S^o{}_s$. It can not be transfered to the electrolyte from $S^o{}_s$, the formation of which is irreversible. This means that the scheme drawn in Fig. 9 is somehow misleading in the sense, that at the very spot in the surface, where $S^{-\cdot}$ is formed, no conduction band (formed of bulk S^{2-}) exists anymore. It only describes a macroscopic behavior.

SUMMARY

Unpinning of band edges at the semiconductor/electrolyte interface is understood as a common phenomenon for n- and p-type materials. Thus, the band edge positions as obtained from flatband potential measurements in the dark, cannot be taken as a fixed value for the interpretation of charge transfer processes. More investigations in this direction are necessary.

REFERENCES

Allongue, P. and Cachet, H. 1985. Band-edge shift and surface charges at illuminated n-GaAs/aqueous electrolyte junctions. *J. Electrochem. Soc.*, 132, 45-51

Heller, A., Leamy, A.J., Miller, B., Johnston, W.D. 1983. *J. Phys. Chem.*, 87, 3239-32

Inoue, T., Watanabe, T., Fujishima, A., Honda, K., Kohayakywa, K. 1977. Suppression of surface dissolution of CdS photoanode by reducing agents. *J. Electochem. Soc.*, 124, 719-722

Kelly, J.J. and Memming, R. 1982. The influence of surface recombination and trapping on the cathodic photocurrent at p-type III-V electrodes. *J. Electrochem. Soc.*, 129, 730-738.

Kühne, H.M. and Tributsch, H. 1986. Energetics and dynamics of the interface of RuS_2 and implications for photoelectrolysis of water. *J. Electroanal. Chem.*, 201, 263-282

Kühne, H. M. (1987). Private communication.

Li, J. L., and Peter, L.M. 1985. The reduction of oxygen at illuminated p-GaP; evidence for a current doubling mechanism. *J. Electroanal. Chem.*, 182, 399-411

McEvoy, A.J., Etman, M. and Memming, R. 1985. Interface charging and intercalation effects on d-band transition metal diselenide photoelectrodes. *J. Electroanal. Chem.*, 190, 225-241.

Meissner, D., Sinn, C., Memming, R., Notten, P.H.L. and Kelly, J.J. 1986. On the nature of the inhibition of electron transfer at illuminated p-type semiconductor electrodes. In *"Homogenous and Heterogenous Photocatalysis"* (Ed. E.Pelizzetti and N.Serpone). (D.Reidel, Dordrecht). pp. 317-333.

Meissner, D., Benndorf, C., Memming, R. 1987a. Photocorrosion of cadmium sulfide: analysis by photoelectron spectroscopy. *Appl. Surf. Sci.*, 1987, 27, 423-436.

Meissner, D., Memming, R., Kastening, B. 1987b. Photoelectrochemistry of cadmium sulfide, part I: reanalysis of photocorrosion and flatband potential. *J. Phys. Chem.*, to be published.

Meissner, D., Lauermann, I., Memming, R., Kastening, B. 1987c. Photoelectrochemistry of cadmium sulfide, part II: influence of surface state charging. *J. Phys. Chem.*, to be published.

Memming, R. 1978. The role of energy levels in semiconductor-electrolyte solar cells. *J. Electrochem. Soc.*, 125, 117-123

Memming, R. and Kelly, J.J. 1981. Electrochemical photovoltaic cells. In *"Photochemical conversion and storage of solar energy"* (Ed. J.S. Connolly). (Academic press, New York). pp. 243-269.

Memming, R. 1987. Charge transfer kinetics at semiconductor electrodes. *Ber. Bunsenges. Phys. Chem.*, 91, 353-361.

Schröder, K. and Memming, R. 1985. Analysis of trapping and recombination effects at n-GaAs. *Ber. Bunsenges. Phys. Chem.*, 89, 385-392.

Tubbesing, K., Meissner, D., Memming, R., Kastening, B. 1986. On the kinetics of electron transfer reactions at illuminated InP electrodes. *J. Electroanal. Chem.* 214, 685-698.

Vanmaekelberg, D., Gomes, W.P. and Cardon, F. 1985. Studies on the n-GaAs photoanode in aqueous electrolytes. *Ber.Bunsenges.Phys.Chem.*, 89, 994-98

H_2O_2 PRODUCTION IN A PHOTOELECTROCHEMICAL CELL WITH TiO_2 ELECTRODES: REACTION MECHANISMS AND EFFICIENCY.

D. Tafalla and P. Salvador

Instituto de Catálisis y Petroleoquímica, C.S.I.C.
Serrano, 119. 28006 MADRID (Spain)

ABSTRACT

The electroreduction of dissolved and/or photogenerated O_2 at a n-TiO_2 electrode was studied. The efficiency of photosynthetic generation of H_2O_2 from water, at both the photoelectrochemical and photocatalytic mode, was measured. A reaction mechanism involving bandgap surface states, probably associated with chemisorbed OH^- groups was proposed. The relative amount of hydrogen peroxide as a final product of O_2 electroreduction was found to be controlled by the rate at which generated H_2O_2 molecules become chemisorbed on the TiO_2 surface giving rise to some kind of titanium peroxo complexes.

INTRODUCTION

The photoproduction of H_2O_2 from water on U.V. irradiated TiO_2 macroelectrodes or aqueous suspensions seems to be an interesting reaction for the conversion of solar energy into fuels (Clechet et al., 1979; Rao et al., 1980; Harbour et al., 1985). Although H_2O_2 is believed to be produced in an intermediate step of the water photoelectrolysis (Salvador, 1985), the main source of H_2O_2 is the electroreduction of photogenerated or dissolved O_2 (Clechet et al., 1979). In fact, at low bandbending, TiO_2 valence band photogenerated holes are able to oxidize water:

$$2H_2O + 4h^+_{VB} \longrightarrow O_2 + 4H^+ \qquad [1]$$

photogenerated oxygen molecules being simultaneously reduced with conduction band electrons:

$$O_2 + 2e^-_{CB} + 2H^+ \longrightarrow H_2O_2 \qquad [2]$$

Using n-TiO_2 electrodes, either naked or platinized, we have studied reaction [2] in a photoelectrochemical cell (PEC) under conditions of controlled electrolyte pH, illumination intensity and external applied bias (Tafalla and Salvador, 1987). These results can be extrapolated to the case of TiO_2 suspensions help to go further into the mechanisms of O_2 photo-uptake.

EXPERIMENTAL

Pollycrystalline TiO_2 electrodes prepared by anodic oxidation of a Ti sheet in saturated NaOH were employed in photocurrent transient measurements. This technique, whose experimental details have been described elsewhere (Salvador, 1985), allows a measurement of the net current of the electroreduction of photogenerated O_2. A 1000 W-Xe lamp was employed as light source, the illumination intensity being controlled with neutral filters. The rotating ring-disk electrode (RRDE) technique was employed to measure the proportion to H_2O_2 generated according to [2] (Tafalla and Salvador, 1987). Electrolyte dissolved O_2 was electroreduced in the TiO_2 disk and the produced H_2O_2 detected in the concentric Pt ring appropriately polarized. From the relationship between the ring and disk current the percent of H_2O_2 produced from the O_2 electroreduction can be obtained.

RESULTS AND DISCUSSION

Cathodic reactions

Under TiO_2 electrode polarization slightly anodic from the flatband potential (V_{fb}), a cathodic current superimposed to the anodic photocurrent (transient behaviour) can be observed (Fig. 1). This catohdic effect is attributed to the recombination with e^-_{CB} of holes trapped at surface species (mainly OH° radicals and H_2O_2 molecules) photogenerated at intermediate steps of oxygen evolution (Salvador, 1985).

These reactions can be written:

$$OH^- + h^+ \longrightarrow OH° \qquad [3]$$
$$OH°_s + e^- \longrightarrow OH^-_s \qquad [4]$$
$$OH°_s + OH°_s \longrightarrow (H_2O_2)_s \qquad [5]$$
$$(H_2O_2)_s + 2e^- \longrightarrow 2OH^-_s \qquad [6]$$

The electroreduction of photogenerated O_2 to water with the participation of four conduction band electrons:

$$O_2 + 4e^- + 4H^+ \longrightarrow 2H_2O \qquad [7]$$

also contributes, together with [2], to the observed cathodic effect.

While OH° and H_2O_2 species interact strongly with the TiO_2 surface, the interaction of photogenerated oxygen is weak, which allows most photogenerated O_2 molecules to be swept away from the electrode surface by simple

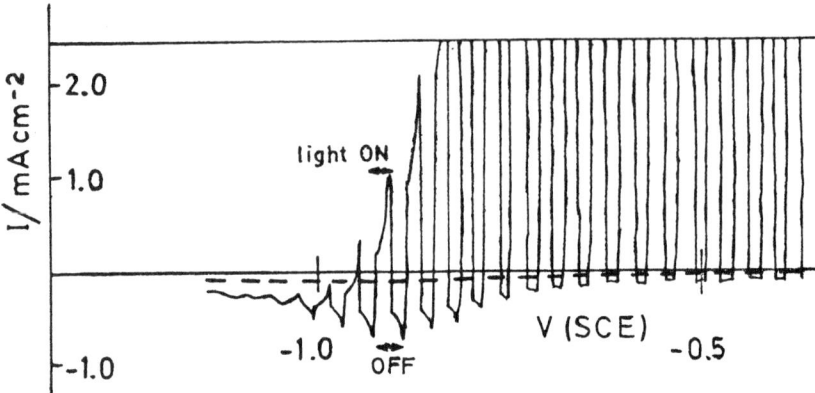

Fig. 1 Detail of the photocurrent-voltage curve under hand chopped light showing the cathodic current superimposed to the anodic photocurrent.

electrolyte stirring. Then, the contribution to the cathodic effect of the electroreduction of photogenerated O_2 can be obtained from the diference between the magnitude of the cathodic transient without and with electrolyte stirring (Tafalla and Salvador, in press). The fact that the overpotential for O_2 electroreduction at TiO_2 in acidic medium is higher than that in basic medium (Tafalla and Salvador, 1987) produces a strong decrease of the reaction 2 at acidic pH.

Rotating Ring-Disk electrode (RRDE) experiments

The RRDE technique allows an estimation of the relative amounts of H_2O_2 (reaction [2]) and H_2O (reaction [7]) produced during O_2 electroreduction. Fig. 2 shows the TiO_2 potential dependence of the H_2O_2 percent obtained as final product of O_2 electroreduction. This percent slowly decrease from 40% down to zero as the bandbending diminishes.

By combining the information from Fig. 2 with that obtained from the cathodic effect shown in Fig. 1, the efficiency of H_2O_2 photoproduction ($\eta_{H_2O_2}$) via the electroreduction of photogenerated oxygen can be estimated. $\eta_{H_2O_2}$ can be defined as the ratio between the H_2O_2 generation rate according to [2] and the photogeneration rate of "effective" e^--h^+ pairs. Fig. 3 shows the dependence of $\eta_{H_2O_2}$ on the applied bias for three different values of the photon flux. A maximum value $\eta_{H_2O_2} \simeq 10\%$ is obtained in the photoelectrochemical mode, whereas under photocatalytic conditions is $\eta_{H_2O_2} \simeq 1\%$ (Tafalla and Salvador, in press).

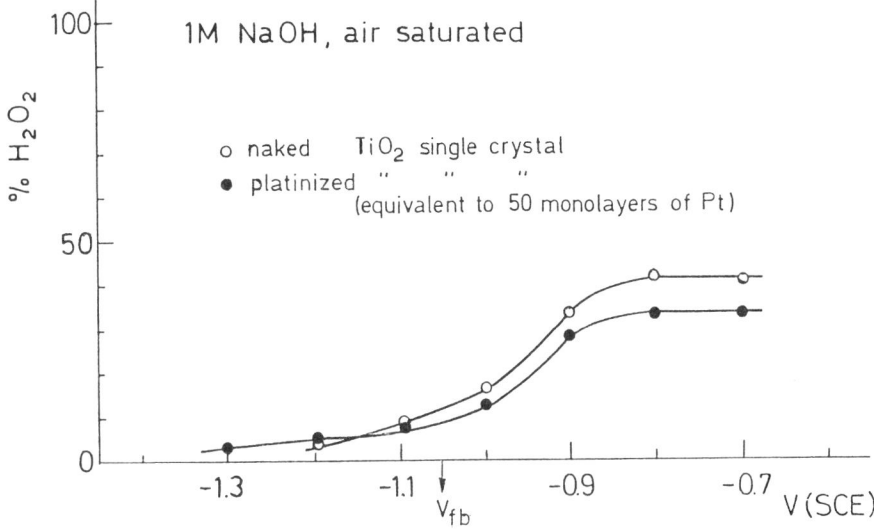

Fig. 2. Disk potential dependence of the percent of H_2O_2 obtained with the RRDE as final product of the electroreduction of dissolved O_2 at a n-TiO_2 single crystal.

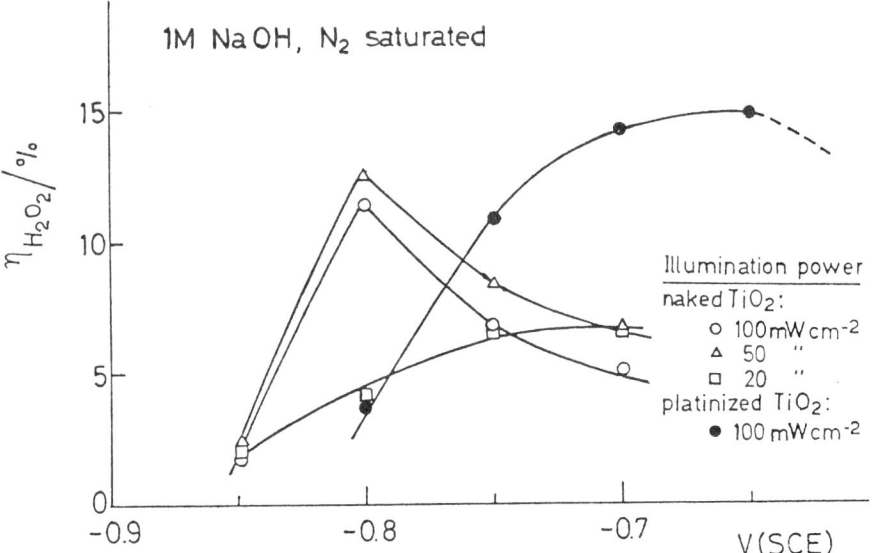

Fig. 3. Efficiency of H_2O_2 photoproduction as a function of the applied potential and illumination intensity.

O_2 electroreduction mechanisms

The following scheme sumarizes the main ways of oxygen electroreduction at a n-TiO_2 electrode:

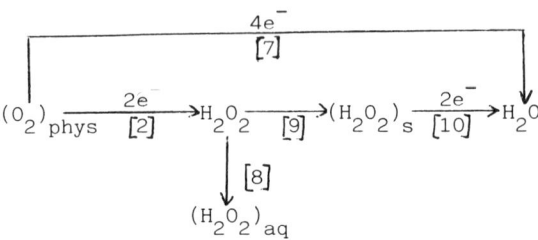

Physisorbed O_2 can be reduced either to H_2O via a four electron reaction or to H_2O_2 (two electron step), which can either diffuse into the electrolyte (step [8]) or be chemisorbed (step [9]) and further reduced to H_2O (step [10]). It is apparent from Fig. 2 that in the low cathodic current region, under polarization anodic of V_{fb}, the rate of step [8] is higher than the rate of [9] and a considerable percent of H_2O_2 can diffuse into the electrolyte and be detected at the Pt ring. As the TiO_2 applied potential becomes cathodic of V_{fb} (high cathodic current region), the rate of [10] exceedes the chemisorption rate and [8] can not compete with [9], so that the amount of H_2O_2 detected at the ring drops.

Tafel plots of O_2 electroreduction at n-TiO_2 in electrolytes saturated with air and O_2, respectively, showed that under electrode polarization anodic of V_{fb} the cathodic current is not proportional to the electrolyte concentration of oxygen (Tafalla and Salvador, 1987). This experimental fact clearly indicates that the mechanism of electron transfer from the semiconductor conduction band to the empty energy levels of dissolved O_2 molecules is not isoenergetic but rather inelastic, involving a two step reaction with the participation of bandgap surface states near the conduction band. On the basis this important result, the following mechanisms of O_2 electroreduction, at basic pH, can be constructed with the help of the energy diagram of Fig. 4:

In a first step conduction band electrons are inelastically captured by bandgap surface states probably associated with chemisorbed OH^- groups. The second, limiting step is the isoenergetic electron transfer from these surface states to dissolved O_2 molecules:

$$O_2 + e^- \longrightarrow O_2^- \tag{11}$$

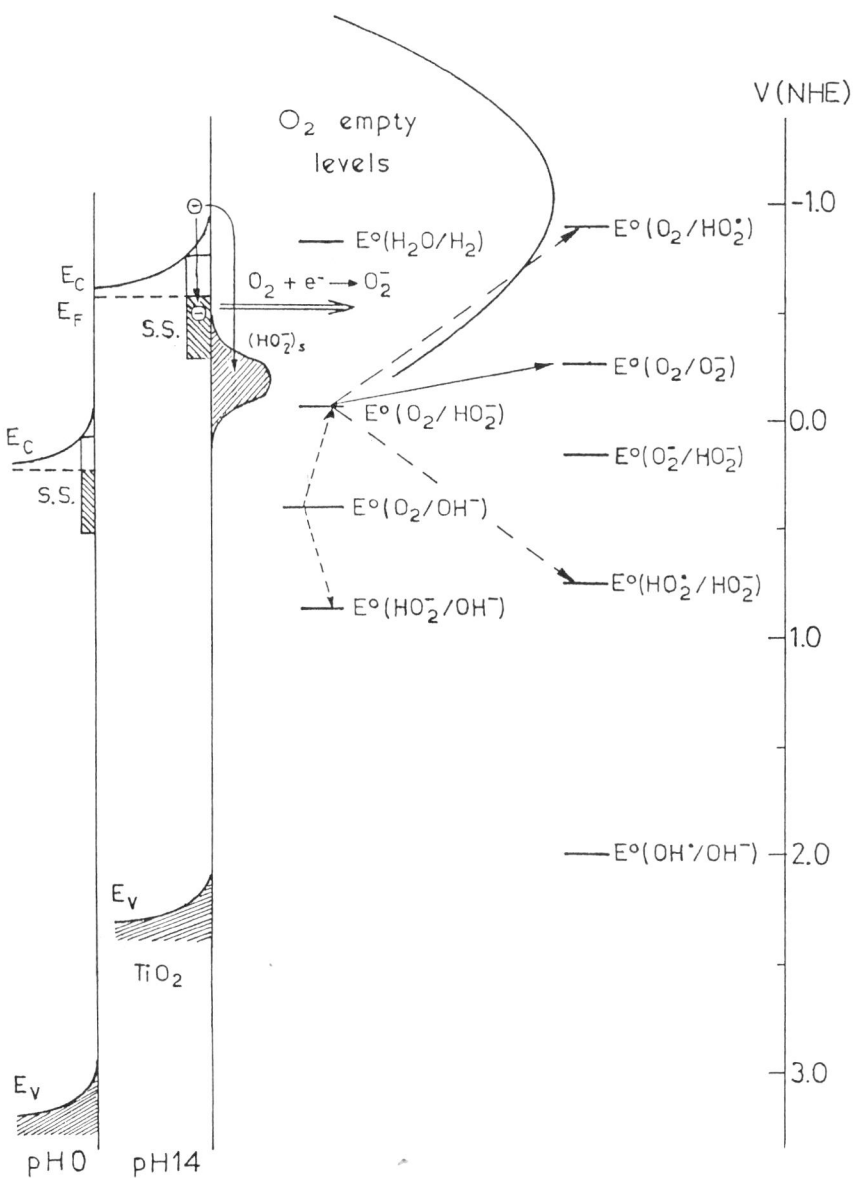

Fig. 4. Energy diagram of the TiO_2-electrolyte interface in basic medium. The TiO_2 energy levels at pH0 are included in order to show the change in overlapping of the surface states (S.S.) with O_2 empty levels.

In a third step the following chemical reaction should takes place:

$$O_2^- + H_2O \longrightarrow HO_2^o + OH^- \qquad [12]$$

and HO_2^o radicals are generated. Further, a second electron is transferred to these radicals from the surface states:

$$HO_2^o + e^- \longrightarrow HO_2^- \qquad [13]$$

This is the last step of reaction [2]. Those HO_2^- species that cannot diffuse into the electrolyte (reaction [8]) become strongly chemisorbed (reaction [9]) and further reduced by inelastic capture of electrons from the conduction band (reaction [10]).

In acidic medium the mechanisms are similar, althoguh the rate of [11] is lower than at basic pH. Firstly, the density of surface states associated with OH_s^- surface groups, is smaller at acidic than at basic pH. Secondly, the overlapping between empty O_2 levels and surface states considerably diminishes because of the shift towards positive potentials of the semiconductor energy levels (including surface states) with respect to the pH independent O_2/O_2^- redox potential (see Fig. 4).

Effect of Platinum overlayers

Pt electrodeposition produces a decreases of the overpotential of O_2 electroreduction and therefore a displacement of the maximum efficiency of H_2O_2 photoproduction toward potentials more anodic (see Fig. 3). However the mechanisms of the reaction seem to be the same as for naked TiO_2 (Tafalla and Salvador, 1987).

REFERENCES

Clechet, P., Martelet, C., Martin, J.R. and Olier, R. 1979. Photoelectrochemical behaviour of TiO2 and formation of hydrogen peroxide. Electrochim. Acta, 24, 457-461.
Harbour, J.R., Tromp, J. and Hair, M.L. 1985. Photogeneration of hydrogen peroxide in aqueous TiO2 dispersions. Canad. J. Chem., 65, 204-208.
Rao, M.V., Rajeshwar, K., Pal Verneker, V.R. and DuBow, J. 1980. Photosynthetic production of H2 and H2O2 on semiconducting oxide grains in aqueous solutions. J. Phys. Chem., 84, 1987-1991.
Salvador, P. 1985. Kinetic approach to the photocurrent transients in water photoelectrolysis at n-TiO2 electrodes. 1. Analysis of the ratio of the instantaneous to steady-state photocurrent. J. Phys. Chem., 89, 3863-3869.
Tafalla, D. and Salvador, P. 1987. Mechanisms of charge transfer at the semiconductor-electrolyte interface: oxygen electroreduction at naked and platinized n-TiO2 electrodes. Ber. Bunsenges. Phys. Chem., 91, 475-479.

Tafalla, D. and Salvador, P. in press. Photosynthetic production of H_2O_2 from water at n-TiO_2 electrodes in a photoelectrochemical cell: influence of electrolyte pH and Pt electrodeposition. J. Electroanal. Chem.

ADSORPTION EXPERIMENTS FOR MODELLING SEMICONDUCTOR/ELECTROLYTE INTERFACES

W. Jaegermann

Hahn-Meitner-Institut Berlin, Bereich Strahlenchemie
1000 Berlin 39, Federal Republic of Germany

ABSTRACT

Adsorption of halogens (Cl_2, Br_2, I_2) and H_2O on semiconducting group VI chalcogenide surfaces and changes due to this adsorption have been investigated related to photoelectrochemical results. Clean (0001) faces are prepared by cleaving the crystals in the UHV and are characterized by UPS, XPS, ISS and LEED. H_2O is molecularly adsorbed inducing band bending on p-type MoS_2 and only a shift of electron affinity on n-type MoS_2. At low doses the photoelectron spectra indicate the halogens are ionsorbed. Changes of the energy correlation at the interface due to the adsorbates are determined from the HeI photoelectron distribution curves. On n-type semiconductors the formation of a depletion layer as well as changes of electron affinity are observed. The induced band bending can be reversed by white light illumination creating a "photopotential", which was found to be dependent on the surface ideality. The values determined for the work function, electron affinity and band bending (approximately equivalent to the observed photopotential) are very similar to values obtained in electrochemical junctions.

In addition the interaction of H_2O and Cl_2 with UHV cleaved (011) faces of n-$CuInSe_2$ have been studied. H_2O adsorbs undissociatively and reversibly and induces no band bending at the interface. Cl_2 adsorbs dissociatively reacting at low temperatures primarily with Cu. After annealing Cu migrates to In forming $InCl_3$. First Fermi level pinning is observed as in electrochemical environment which is reduced after $InCl_3$ formation leading to a small photopotential.

The possible results and limitations of model experiments for semiconductor/electrolyte interfaces are discussed for non-reactive and reactive interfaces and related to the use of UHV techniques for obtaining microscopic information of interfacial processes in photoelectrochemical processes.

INTRODUCTION

The electronic properties of semiconductors junctions are strongly dependent on their interfaces. This is especially true for semiconductor/electrolyte contacts as in photoelectrochemical solar cells, for which a variety of possible reactants must be considered.

Electronic and structural changes of the semiconductor surfaces may influence light energy conversion yields or catalytic enhancements of desired reactions to chemical fuels. The involved processes are up till now not really understood on a molecular level.

Therefore it is very important to complete the data obtained by (photo)electrochemical techniques with surface sensitive spectroscopic measurements. One promising possibility of gaining microscopic information on interfacial processes is the use of UHV surface science techniques. However due to the analysis requirements emersion of the samples from the electrolyte and transfer into UHV is necessary. During this procedure the semiconductor interface may change drastically. Alternatively the basic chemical and physical interactions of electrolyte components may be studied by adsorbing redox components on defined semiconductor surfaces thus simulating semiconductor/electrolyte junctions.

In this paper our results to simulate the photoactive semiconductor/electrolyte interface in UHV by adsorbing halogens and H_2O on semiconductor surfaces are described. For these experiments layer type compounds and ternary chalcogenides have been considered because clean faces can easily be prepared by cleaving the crystals in UHV and because the reactions with halogens are intensively studied for photoelectrochemical solar cells.

The cleaved surfaces are characterized by UPS, XPS, LEED and ISS. The adsorbate induced changes are followed by ISS, UPS and XPS, which allow the investigation of adsorbed species and the investigation of their electronic influence.

THEORETICAL BACKGROUND

For samples in good ohmic contact to the detector system the photoelectron energy distribution curve is referred to the Fermi level E_f. Adsorbate induced shifts of the photoemission spectra are thus related to changes of the binding energy values E_B and changes of the work function ϕ:

$$E_B = h\nu - E_{kin} - \phi \quad \text{and} \quad \Delta\phi = -eV_{bb} + \Delta\chi$$

The band bending eV_{bb} at n-type semiconductors is expected to decrease E_B and additional changes of ϕ are related to changes of the electron affinity χ (fig 1). As the band bending can possible be reserved by illumination a light induced opposite shift of the spectrum equivalent to a photopotential (U_{ph}) is expected. The relation between the electrochemical energy scale $E_{red/ox}$ vs NHE and the absolute energy scale E vs vacuum level is made by the relation:

$$E_{red/ox} = -4,5 - E.$$

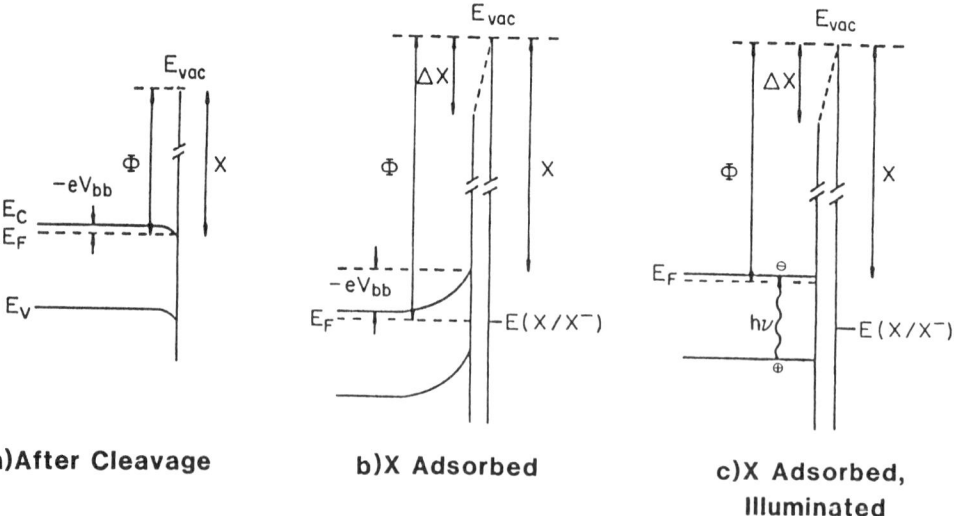

a) After Cleavage   b) X Adsorbed   c) X Adsorbed, Illuminated

Fig. 1 Schematic energy diagram of the semiconductor/adsorbate interface for different experimental conditions.

EXPERIMENTAL

Single crystalline n-CuInSe$_2$ was grown from the melt by the gradient freeze technique.

Single crystals of CuInS$_2$ and the layered semiconductors were prepared by chemical transport reactions. Iodine transported n-type material had the best electronic properties among the CVT grown samples and was used here.

Surface analytical experiments were carried out in a commercial UHV system (Escalab Mk II, Vacuum Generators). The base pressure during experimentation was in the 10^{-11} mbar range. Excitation sources were Mg Kα (1253,6 eV) for XPS and HeI (21,2 eV) for UPS measurements. For experimentation the crystals were cleaved in UHV. LEED experiments were performed with a four grid reverse view LEED system (VSW). For ISS measurements a low energy ion source (VG, AG61) was used.

For the adsorption experiments the gases were introduced through a specifically designed dosage nozzle. The gases were adsorbed after cooling the samples to liquid N$_2$ (LN$_2$) temperatures. A more detailed description of the system and the experimental procedure is given elsewhere (Jaegermann, 1986).

EXPERIMENTAL RESULTS

The interaction of H$_2$O and halogens with clean single crystalline (0001) van der Waals faces of MoX$_2$ (X= S, Se) have been studied as an example for the behaviour of ideal non-reacting interfaces.

A well defined LEED pattern of hexagonal symmetry and IS-spectra showing no surface contaminations were obtained for the clean surfaces.

Adsorption of H_2O on MoS_2 only occurs for samples cooled to LN_2-temperatures (Jaegermann and Schmeißer, 1987). The adsorbed H_2O shows an emission pattern of three broad bands, centered at 7,10 and 13 eV, which is typical for physisorbed H_2O. On n-type samples a shift of the MoS_2 related emission peaks is not observed (Fig. 2 a). A decrease of ϕ by 0.3 eV is related to the change of χ due to surface dipoles. For p-type samples, how-ever, adsorption of H_2O leads to a downward bending of semi- conductor bands, which increases the measured values for E_B and ϕ by the same amount (0.5 eV) (Fig. 2 b). This indicates possible charge transfer from the electron donating H_2O into the semiconductor, which is only ex- pected for p-type and not for n-type semiconductors. A reaction of H_2O with the surface does not occur. By warming up the sample to room tempera- ture H_2O completely desorbs restoring the original emission pattern.

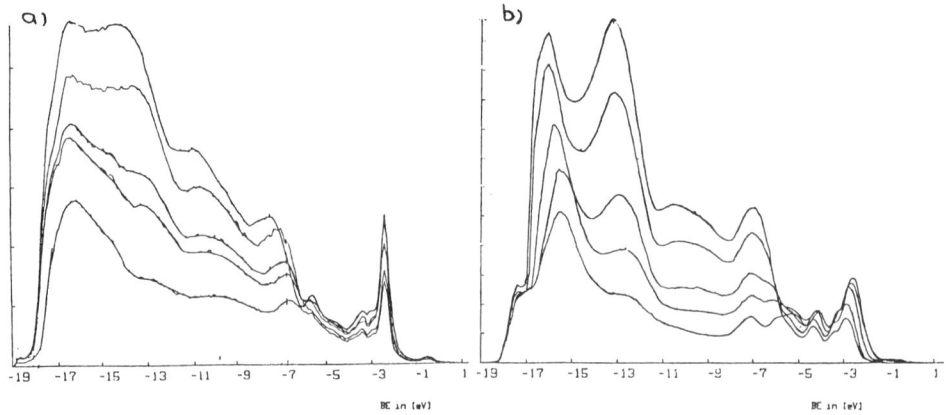

Fig. 2 HeI photoelectron spectra of n-type a) and p-type (b) MoS_2 (0001) for increasing coverages by H_2O.

Adsorption experiments of halogens as typical electron acceptors on n-MoSe$_2$(0001) indicate a reverse behavior with respect to possible charge transfer (Jaegermann, 1986; Jaegermann and Schmeißer, 1987). Due to electron flow from the n-type semiconductor to the adsorbed halogens a decrease of E_B indicating an upward bending of energy band occurs. In addition ϕ increases by an even larger amount (compare Fig. 3 for Cl$_2$ adsorption).

The shift of E_B can partly be removed by illumination of the sample with white light (W Hal lamp, 40 mW/cm²) (Fig. 4). As the observed surface photovoltage is considerably smaller than the shift of E_B due to adsorption the formation of an inversion layer on this small bandgap semiconductor has to be assumed. The energy correlation at the semiconductor/adsorbate interface is shown in Fig. 1. It corresponds to the diagram of the semiconductor/electrolyte interface as is suggested by a comparison of contact potential differences and photopotentials obtained for the different halogens in UHV and in electrochemical junctions (organic electrolytes) (compare Table 1).

A drastic decrease of photovoltage in UHV is obtained by introduction of surface states at the semiconductor surface. Particle bombardement of cleaved (0001) faces leads to preferential sputtering of the chalcogenide. The metal is reduced and new electronic bandgap states are formed at the surface. As a consequence a Fermi level pinning effect occurs which results in a smaller shift of E_B due to halogen adsorption and decreased photovoltages and consequently an increased double layer potential drop (Fig. 4).

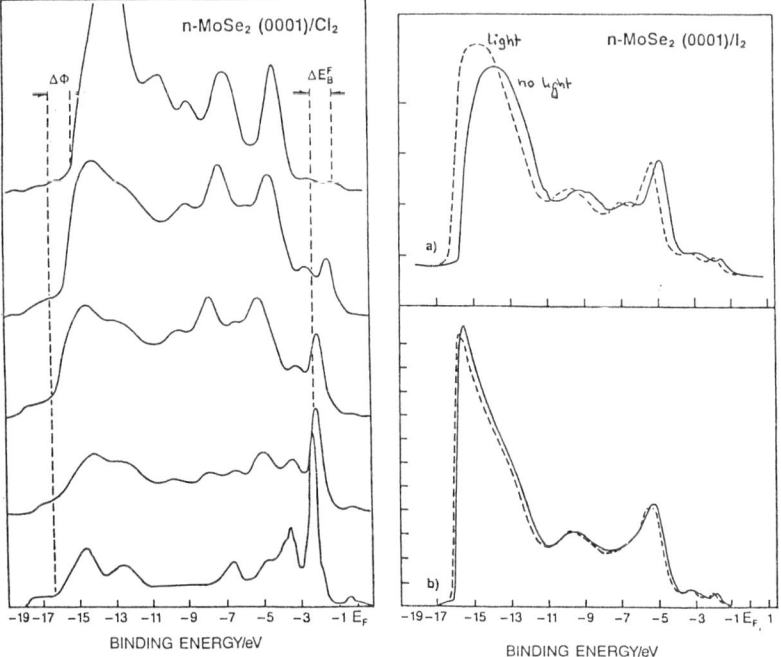

Fig. 3 HeI photoelectron spectra of n-MoSe$_2$ (0001) for increasing coverage by Cl$_2$

Fig. 4 HeI photoelectron spectra of n-MoSe$_2$ (0001) a) and sputtered n-MoSe$_2$ b) with 1 ML of I$_2$

Table I   Comparison of energy correlations in UHV adsorption experiments (shift of binding energy ΔE_B^F, shift of work function $\Delta\phi$, and surface photovoltage U_{PH}) and at electrochemical interfaces (standard redox potentials E_o and photopotential U'_{PH})

	ΔE_B^F [eV][a]	$\Delta\phi$ [eV][a]	U_{PH} [eV]	E_0 [V][b]	U'_{PH} [V][c]
Cl$_2$	1.4	1.8	0.7	1.4	0.7
Br$_2$	1.2	1.8	0.7	1.1	0.5
I$_2$	0.5	0.9	0.5	0.5	0.3

a) related to ϕ = 4.6 eV corresponding to 0.1 V vs. NHE
b) vs. NHE
c) related to U_{fb} = 0.25 V vs NHE, mean value

The XP-spectra indicate that a reaction of the adsorbed halogens with the semiconductor surface does not occur. There are no extra emission bands or broading of the substrate peak widths which would be expected for possible reaction products. In addition after desorption of the halogens the original emission patterns of the UP- and XP-spectra are reestablished showing complete desorption of the adsorbed halogens. The halogens are molecularly adsorbed for higher doses as is inferred from the emission pattern especially in UPS. The observed charge transfer suggests a partial adsorption as ionized species which, however, due to the small number involved (1 % to 10 % of a monolayer) was not detected spectroscopically.

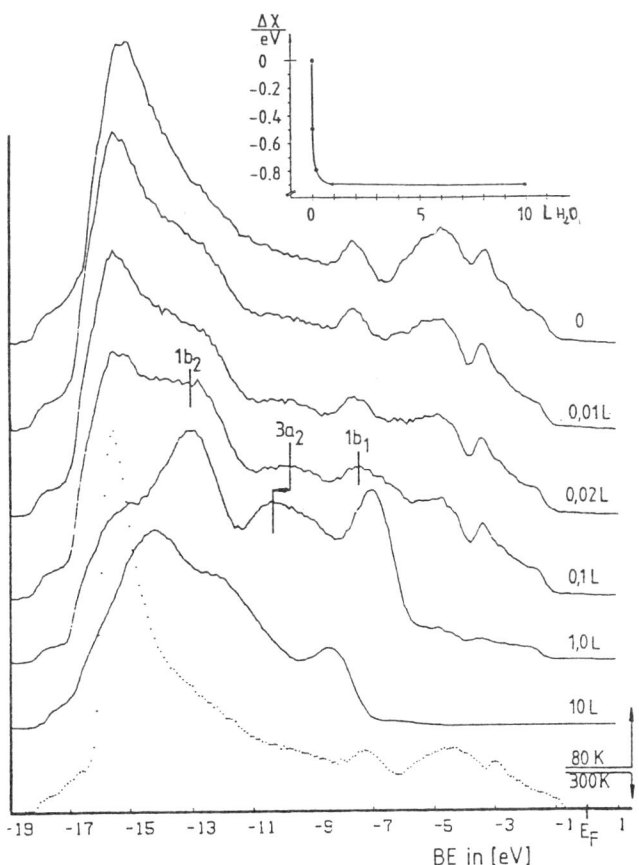

Fig. 5 HeI photoelectron spectra of n-CuInSe$_2$ for increasing coverage by H$_2$O and after annealing. The insert shows the change of electron affinity.

The good correspondence of semiconductor/adsorbate interfaces to electrochemical junctions for non-reacting interfaces immediately leads to the question how reactive interfaces may behave. For this purpose H$_2$O and Cl$_2$ adsorption on cleaved CuInX$_2$ (X = S, Se) surfaces have been investigated. The cleaved CuInX$_2$ surfaces show the rectangular LEED pattern expected for the (011) face. Electrochemical investigations clearly indicate a reactive transformation of the surface during electrochemical treatment (Bachmann, 1984).

Fig. 5 shows UP-spectra obtained on clean and water covered n-CuInSe$_2$ surfaces (Sander et al, 1987). Compared to the clean surface, a threefold structure between 14 and 7 eV develops upon water dosage at LN$_2$ temperatures. Upon heating to 300 K, the original spectrum is recorded again indicating complete desorption of H$_2$O.

A simultaneous work function change upon water dosage is also displayed in Fig. 5. It is seen that the most pronounced changes are introduced in the low coverage regime well below 0.2 L (1 L $\approx$ 1 · 10^{-6} mbar · s) exposure. The change itself is quite pronounced resulting in a saturation near $\Delta\chi \approx$ 0,9 eV at exposure > 1 L. A shift of the spectral features due to band bending is not observed. In the XP-spectra in addition to the unchanged substrate peak a O(1s) peak shows up at E_B = 533 eV. These results indicate that H$_2$O is physisorbed on CuInX$_2$ with the O directed towards the surface for low coverages. At high coverages hydrogen bonds are formed, which leads to the formation of adsorbed H$_2$O clusters (ice). A reaction at the interface does not occur.

However, adsorption of Cl_2 leads to drastic changes of the surface (Fig. 6,7) (Sander et al., 1987). For low coverages (< 1L) the UP- and XP-spectra show the formation of dissociatively adsorbed Cl_2 (Cl^-). At higher coverages also molecularly adsorbed Cl_2 is detected. After annealing the samples at 300 K the spectra of the clean faces are not reproduced and Cl^- is still present at the surface. The substrate XP-emission peaks indicate that Cl^- originally interacts with Cu sites. At 300 K a slow migration of Cl^- from the Cu sites to In sites occurs and $InCl_3$ is formed.

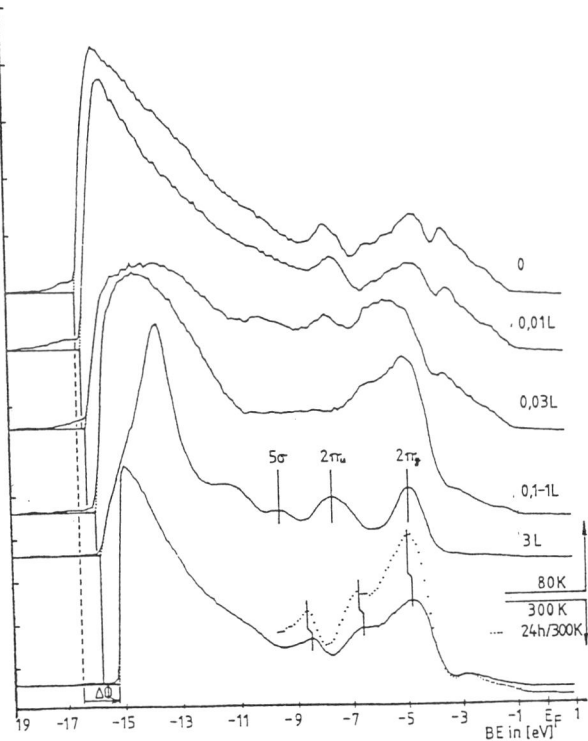

Fig. 6 HeI photoelectron spectrum of n-$CuInSe_2$ for increasing coverage by Cl_2 and after annealing.

As a consequence of the complicated surface chemistry the energetic conditions of the interface do not correspond to that of ideal semiconductor junctions following Schottky's theory. The overall binding energy shift of the spectra is smaller than expected (0.2 eV) and only small photovoltages are obtained. Evidently the surface reaction products form new elec-tronic surface states in the bandgap of the semiconductor (may be Cu^{2+}) These results resemble the situation of $CuInSe_2$ in Cl^- containing electrolytes in which a reactive surface passivation is observed.

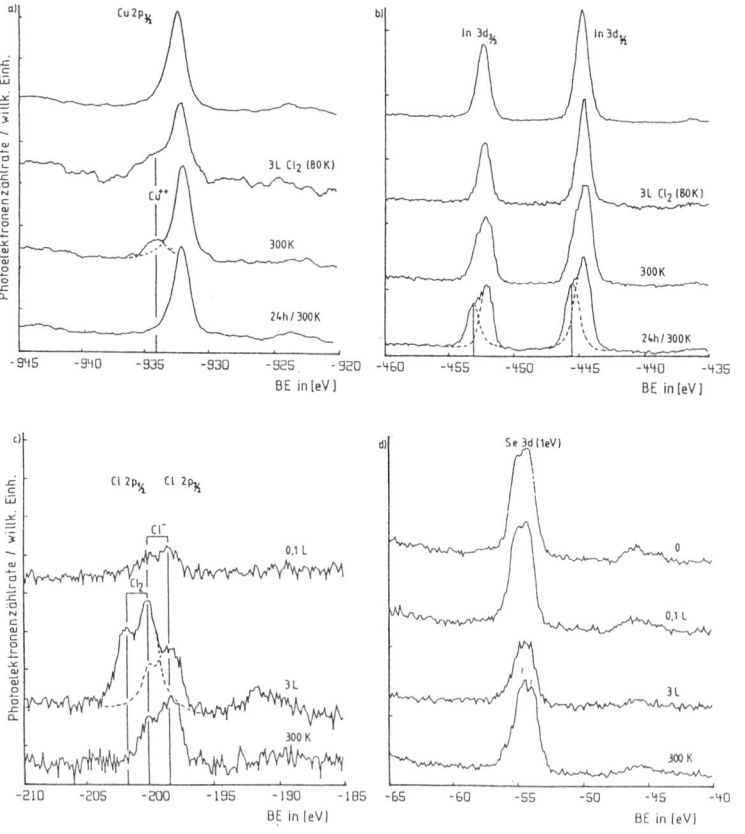

Fig. 7 X-ray photoelectron spectra of n-$CuInSe_2$ of clean surfaces, Cl_2 covered surfaces and annealed surfaces.

After transformation of the surface due to annealing a small photoeffect of 0.4 eV develops, which is related to the partial removal of Cu^{2+} and the formation of insulating $InCl_3$ reducing the effect of Fermi level pinning.

DISCUSSION AND CONCLUSIONS

The above presented results indicate that adsorption experiments of electrolyte components on defined semieconductor surfaces may help to develop a microscopic understanding of interfacial processes. It is principally possible to obtain spectroscopic information on the type of chemical species adsorbed on the surface and in addition the consequences for the energy diagram can be derived from measurements of work function ϕ, binding energy shift ΔE_B and surface photovoltage U_{ph}. In the case of inert surfaces similar conditions as in electrochemical junctions are obtained as shown for the n-$MoSe_2$/halogen system. The situation for reactive interfaces is more complex. The reactivity of the surface and the primary reaction products may be deduced from adsorption experiments as presented here. Also in these cases the important parameters of the enery diagram are attainable, which again show a similar qualitative behaviour as in electrochemical junctions (Fermi level pinning for $CuInSe_2/Cl_2$). However, it must be considered that in electrochemical environment the dissolution of reaction products may be an important step in the transformation of the interface which is not possible in UHV.

Summarizing the possible use of UHV surface science techniques for photoelectrochemical systems it can be concluded that only a combined approach of UHV measurements may lead to a complete microscopic understanding of interfacial properties. The chemical analysis of semiconductor surfaces after a sequence of chemical and/or electrochemical treatments will give important informations on chemical transformations of the interface. The analysis is mostly restricted to XPS and Auger spectroscopy due to problems with surface contamination. An optimized transfer system, which has to be developed may extend the number of possible techniques. Model experiments, as presented here, which, in addition, allow the investigation of changes in the energy scheme in dependence on surface interactions may help to bridge the gap between results obtained by sensitive but unspecific electrochemical techniques and often unproven interpretations based on chemical intuition.

REFERENCES

Bachmann K.J., Menezes S., Kötz R., Fearheiley M. and Lewerenz H.J. 1984, Surf. Sci. 138, 457

Jaegermann W. 1986, Chem. Phys. Lett. 126, 301

Jaegermann W., and Schmeißer D. 1987, J. Vac. Soc. A 5, 627

Sander M., Lewerenz H.J., Jaegermann W. and Schmeißer D., 1987, Ber. Bunsenges. Phys. Chem. 91, 416

ACKNOWLEDGEMENTS

Gratefully the generous support by and stimulating interaction with Prof. Tributsch is acknowledged. I have to thank my colleagues especially D. Schmeißer and M. Sander who are involved in parts of the work presented here and who are cited in the references.

CYTOCHROME b-559 AS A TRANSDUCER OF REDOX ENERGY INTO ACID-BASE ENERGY IN PHOTOSYNTHESIS

M. Losada, M. Hervás and J.M. Ortega

Instituto de Bioquímica Vegetal y Fotosíntesis,
Consejo Superior de Investigaciones Científicas y Facultad de
Biología, Apartado 1113, 41080 Sevilla, Spain

ABSTRACT

The redox and acid-base states as well as the midpoint potentials and pK_a's of cytochrome b-559 have been determined in intact thylakoids and oxygen-evolving PS II particles (which are devoid of cytochromes f and b-563) at room temperature in the pH range from 6.5 to 8.5. At pH 7.5, the native cytochrome is about 2/3 in its reduced and protonated (nonautooxidizable) high-potential (HP) form and about 1/3 in its oxidized and nonprotonated low-potential (LP) form. The protonated HP couple is pH-independent ($E'_o \simeq +0.36$ V), its oxidized form being electronically energized and labile, whereas the nonprotonated LP couple is stable and pH-independent above pH 7.6 ($E'_o \simeq +0.12$ V), but becomes pH-dependent with a slope of about -60 mV per pH unit below this pH. Cytochrome b-559 is located in the photosynthetic noncyclic electron transport chain between the reducing side of PS II and the oxidizing side of PS I, and there is a close relationship between light-induced proton translocation coupled to electron transport and the content of HP cytochrome b-559 in thylakoids. It is concluded that cytochrome b-559 functions, as a particular case of a general model, as a transducer of redox energy into acid-base energy by operating at two alternate E'_o's and two alternate pK_a's.

INTRODUCTION

During the photosynthetic process (Whatley and Losada, 1964; Hill, 1965; Arnon, 1977; Boyer et al., 1977; Losada and Guerrero, 1979; Losada et al., 1987) there is a sequential transduction of light energy into redox energy, acid-base energy, metaphosphate-orthophosphate energy and chemical-bond energy. In other words, photons energize electrons; subsequently, electrons energize protons; protons, in turn, energize orthophosphate (P_i) to metaphosphate ($\sim P^{**}$), and, finally, metaphosphate energizes a variety of chemical groups (X to $\sim X^{**}$), such as carboxylate anions to acylium cations, sulfate anion to sulfurylium "zwitterion", sugars to glycosyl carbocations, etc, as diagrammatically represented by the following scheme:

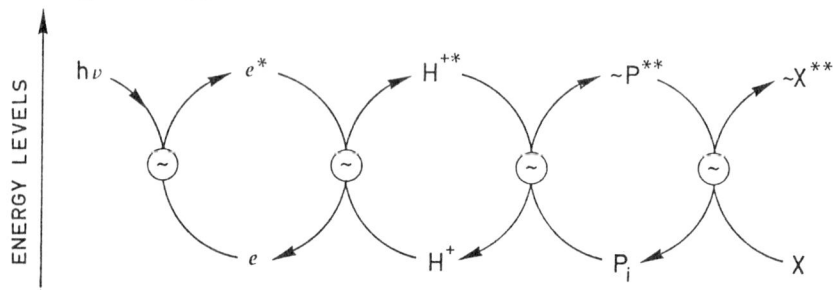

Electronic energy -resulting either from electron excitation or localization- seems to be the compulsory link between the different kinds of energy being transduced by these biochemical systems. All these energy-transducing systems seem to operate -depending on their nature and character of the energization- at two alternate midpoint redox potentials, at two pK_a's or at two phosphate transfer potentials, the key in energy coupling between any two of them being apparently that they share a common intermediate that cyclically participates in the overall transduction process by alternating between its electronically energized state and its unenergized basal state (Losada, 1978; Losada et al., 1983; Losada, 1985; Losada, 1986).

Transduction of light energy into redox energy

The great discovery of photosynthesis was the introduction in cell metabolism of 1-electron redox photosystems, namely, chlorophyll a reaction centers, obviously the most important and universal light-redox energy-transducing systems in biology. They are functionally characterized by being able to operate at two alternate midpoint redox potentials: the high-potential pair, whose reduced form is unenergized (chl), and the low-potential pair, whose reduced form is energized by a photon and electronically excited (chl*), both redox couples sharing the same oxidized form (chl^+):

$$chl^* \xrightleftharpoons{LE'_o} chl^+ + e^*$$

$$chl \xrightleftharpoons{HE'_o} chl^+ + e$$

By operating in this way, the chlorophyll a photocenters of PS II and PS I, acting in series across the thylakoid membrane, can promote the light-driven uphill electron transfer from appropriate high-potential (e) electron donors -ultimately, water- to suitable low-potential (e*) electron acceptors -ultimately, the oxidized primordial bioelements- to a potential difference of about 1 V:

Thylakoid membrane: $\quad\quad\quad\quad e \xrightarrow{2\ h\nu} e^*$

Transduction of redox energy into acid-base energy

In the photosynthetic apparatus, the oxidation of water at high-potential (e) occurs inside the thylakoid vesicles and is concomitant with the liberation of stoichiometric amounts of molecular oxygen and protons at low pH (H_i^{+*}), whereas the reduction of the primordial bioelements (A) at low potential (e*) takes place outside the thylakoids and is concomitant with the fixation of stoichiometric amounts of protons at high pH (H_o^+). As a consequence, the overall redox/acid-base process brings about, besides the uphill electron transfer, the liberation and fixation on either side of the thylakoid membrane, respectively, of one proton per electron transferred at the expense of two photons acting sequentially through the cooperation of the two photosystems, according to the following equations:

Intrathylakoid space: $\quad\quad 2\ H_2O \longrightarrow 4\ e + O_2 + 4\ H_i^{+*}$
Extrathylakoid space: $\quad\quad A + 4\ e^* + 4\ H_o^+ \longrightarrow AH_4$

Total: $\quad\quad\quad\quad\quad 2\ H_2O + A + 4\ H_o^+ \xrightarrow{8\ h\nu} AH_4 + O_2 + 4\ H_i^{+*}$

A very important additional fact, which constitutes the main object of the present article, is concerned with the intermediate downhill transfer of electrons from the reducing side of PS II to the oxidized side of PS I. The corresponding potential gap of about +0.4 V allows, with a remarkable sense of biological efficiency, the translocation from the outside to the inside of thylakoids of one supplementary proton per electron transferred, according to the overall equation:

Thylakoid membrane: $\quad 4\ e^* + 4\ H_o^+ \longrightarrow 4\ e + 4\ H_i^{+*}$

Thus, light energy is eventually transduced not only into redox energy, but also into acid-base energy:

$$2\ H_2O + A + 8\ H_o^+ \xrightarrow{8\ h\nu} AH_4 + 8\ H_i^{+*} + O_2$$

REDOX AND ACID-BASE CHARACTERIZATION OF CYTOCHROME b-559 IN THYLAKOIDS AND PS II PARTICLES

Both spinach thylakoids and oxygen-evolving PS II particles contain in their native state about 2/3 of their cytochrome b-559 in its reduced HP form, that is oxidizable by ferricyanide and reducible by hydroquinone, and the remaining 1/3 in its oxidized LP form, that is reducible by dithionite. The absence of the other two redox forms proper to the two redox couples can be explained for the oxidized HP form is energized and very unstable, and the reduced LP form is autooxidizable, both species being, thus, easily transformed in the same oxidized and stable LP form.

Potentiometric reductive titration, using both fresh thylakoids and PS II particles previously oxidized with ferricyanide, has revealed that the LP couple exhibits a constant midpoint redox potential ($\underline{E}'_o \simeq +0.12$ V) above pH 7.6, but becomes pH-dependent below this pH, with a slope of about -60 mV per pH unit, whereas the HP couple is pH-independent in the pH range between 6.5 an 8.5 ($\underline{E}'_o \simeq +0.36$ V). In general, cytochrome b-559 exhibits potential values about 40 mV lower in thylakoids than in PS II particles. After mild heating of the fresh preparations or treatment with the detergent Triton X-100, the HP couple is converted into the LP couple, which preserves its characteristic pH-dependence. In contrast, in the presence of the uncoupler CCCP, the HP couple is also converted into the LP couple, but the pH-dependence proper to the latter is now lost.

The reduced HP form initially present in thylakoids and PS

II particles can, indeed, either be oxidized directly by ferricyanide in the presence of Triton X-100 or become autooxidizable through the action of CCCP. In both cases, the resulting oxidized species is eventually the stable and nonprotonated LP form, but the process follows in either case an indirect and different path. Actually, the protonophoric uncoupler CCCP removes the proton from the reduced and protonated HP form and allows autooxidation at low potential of the resulting deprotonated LP form to its corresponding oxidized LP form. On the other hand, oxidation of the reduced and protonated HP form by ferricyanide after treatment with Triton X-100 occurs compulsorily at high potential, but the resulting protonated HP form is very labile in membranes altered by the action of the detergent and immediately deenergizes itself by proton dissociation to the stable LP form.

Endergonic conversion of the oxidized LP form into the oxidized HP form has been remarkably achieved in PS II particles in accordance with the established redox, acid-base and energetic properties of cytochrome b-559 (Galván et al, 1983; Losada et al., 1983; Hervás et al., 1985; Ortega et al., 1987).

LOCATION AND FUNCTION OF CYTOCHROME b-559 IN THE CHLOROPLAST NONCYCLIC ELECTRON TRANSPORT CHAIN

Our results (Hervás et al., 1986; Ortega et al., 1987) strongly support cytochrome b-559 location in the noncyclic electron transport chain between the reducing side of PS II (plastoquinone) and the oxidizing side of PS I (cytochrome f). Red light (650 nm) absorbed by photosystem II induces reduction of cytochrome b-559 in fresh thylakoids and PS II particles, the photoreduction being completely inhibited by DCMU. In the first case, subsequent illumination with far red light (720 nm) absorbed by photosystem I causes reoxidation of the cytochrome, the rate and magnitude of the photooxidation reaction being greatly enhanced in the presence of 1 µM CCCP.

A close relationship exists (correlation coefficient = 0.98) between light-induced proton translocation coupled to noncyclic electron flow, with either $NADP^+$ or ferricyanide as Hill reagent, and the percentage of cytochrome b-559 in its HP form.

MECHANISM OF ENERGY-COUPLING BY CYTOCHROME b-559

Cytochrome b-559 seems to function as a redox/acid-base transducer between the two photosystems by operating at two alternate E'_o's and pK_a's. Apparently, the LP couple is unprotonated, its reduced form being autooxidizable, and its oxidized form energetically stable. By contrast, the HP couple seems to be protonated, its reduced form being nonautooxidizable and its oxidized form electronically energized and labile:

$$\text{ferrocyt. b} \xrightleftharpoons{LE'_o} \text{ferricyt. b} + e*$$

$$H^+\text{-ferrocyt. b} \xrightleftharpoons{HE'_o} H^+\text{-ferricyt. b*} + e$$

The reduced forms of both redox couples are isoenergetic species of an acid-base pair with a pK_a group of 7.6 (HpK_a). The oxidized forms of the two redox couples constitute likewise an acid-base pair, but of low pK_a (LpK_a), the protonated HP form being an electronically energized species with respect to the unprotonated LP form. Since the potential difference between the two redox couples is about 240 mV, it can be estimated that the pK_a of the oxidized pair can be as low as 3.6, or rather higher, up to 5.6, depending on the role played by the electrical potential across the thylakoid membranes in the energy-transducing process:

$$H^+\text{-ferrocyt. b} \xrightleftharpoons{HpK_a} \text{ferrocyt. b} + H^+$$

$$\text{ferricyt. b*}-H^+ \xrightleftharpoons{LpK_a} \text{ferricyt. b} + H^{+*}$$

Both systems, thus, share as a common intermediate the same oxidized-basic form, which can exist either in its stabilized basal state (ferricyt. b) or in its electronically energized state (ferricyt. b*). Consequently, cytochrome b-559 belongs to the energy-transducing systems of both the energized oxidized-form type and the energized basic-form type, which can reversibly transduce redox energy into acid-base energy. Energization of the oxidized-basic form by the redox system can be coupled with its deenergetization by the acid-base system, as diagrammatically represented by the following scheme:

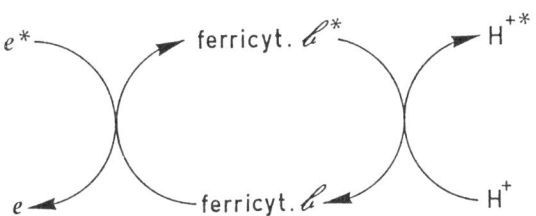

The redox properties of cytochrome b-559 have also been characterized recently by electron paramagnetic resonance spectroscopy in thylakoid membranes, oxygen-evolving PS II particles and the purified protein (Bergström and Vänngard, 1982; Babcock et al., 1985; Ghanotakis et al., 1986). The oxidized LP and HP forms have been identified as low-spin hemes with g_z values of 2.94 and 3.08, respectively, these signals disappearing upon reduction. The heme iron is apparently perpendicular to the membrane plane and ligated in its axial positions by histidine nitrogens from two polypeptide chains. For the LP form, the two imidazol rings are parallel to each other, and there is a large differential effect on the energies of the iron d_{xz} and d_{yz} orbitals (large ligand field splitting), the rhombicity parameter being also large (0.58). For the HP form, a more perpendicular alignment of the imidazol planes occurs, and the differential effect on the d_{xz} and d_{yz} orbital decreases (small ligand field splitting), as does the rhombicity parameter (0.41). Thus a simple shift in the histidines geometry (presumably controlled by redox and acid-base reactions in the membrane environment) is related to the ESR properties of cytochrome b-559.

Taking into account all the preceding considerations, we have further developed our previous models (Losada, 1978; Galván et al., 1983; Losada et al., 1983; Hervás et al., 1985; Losada, 1986; Losada et al., 1987; Ortega et al., 1987) in

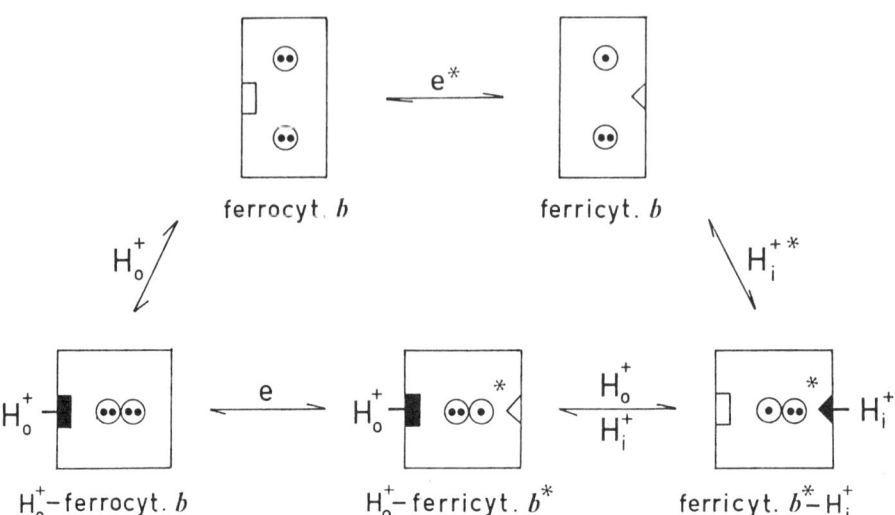

Fig. 1. Diagrammatic representation of the reversible transduction of redox energy into acid-base energy by the cytochrome b-559 system (Losada et al., 1987; Ortega et al., 1987).

order to explain accordingly the role of cytochrome b-559 as a transducing system of redox energy into electronic energy and acid-base energy that operates at two alternate midpoint potentials between the two photosystems and at two alternate pK$_a$'s between the external and internal thylakoid spaces (Fig. 1). The oxidized and nonprotonated LP form of the cytochrome (ferricyt. b) accepts at low potential one electron (e*) in a semifilled vertical-planar d_z orbital and becomes reduced (ferrocyt. b). The reduced LP form changes isoenergetically its electronic configuration upon fixation of one proton (H_o^+) to a high pK$_a$ group at the membrane outside of the thylakoid, thus becoming transformed into the reduced and protonated form (H_o^+-ferrocyt. b) of the HP couple. The oxidation of this form is compulsory at high potential by removal of one electron (e) from the other vertical-planar d_z orbital involved in the redox process, thus becoming electronically energized (H_o^+-ferricyt. b*). The resulting oxidized and protonated HP form undergoes thereafter the displacement of one electron from the filled d_z orbital to the other isoenergetic semifilled d_z orbital, a process that is concomitant with translocation of the proton

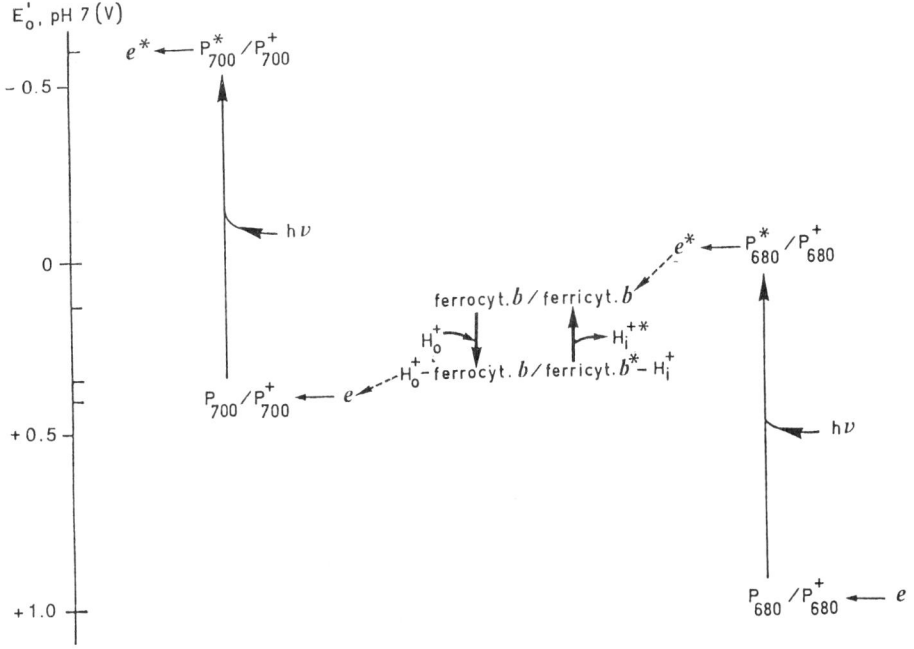

Fig. 2. Diagrammatic representation of the energy-transducing steps involved in the photosynthetic noncyclic electron flow.

from the high pK_a group at the outside of the membrane to another high pK_a group at the inside (ferricyt.$\underline{b}$*-H_i^+). Finally, this oxidized and protonated HP species deenergizes itself to the original oxidized and unprotonated LP form by dissociating its proton at low pH towards the thylakoid lumen (H_i^{+*}), a process that is again accompanied by a change in electronic configuration of the heme iron, but, in this case, to a more stable configuration, the resulting LP form exhibiting thus a lower pK_a group.

In summary, there is general agreement that the uphill noncyclic flow of electrons from water to ferredoxin requires the cooperation of two photosystems operating in series at the expense of visible light and that the generated redox potential gap between both photosystems allows proton translocation against an electrochemical proton gradient across the thylakoid membrane. Regarding the mechanisms of the energy-transducing steps involved in this photosynthetic process, it is our conclusion that pigments P_{680} and P_{700} of photosystems II and I, respectively, both of which transduce light energy into redox energy, belong to the energized reduced-form type, whereas cytochrome b-559, which transduces redox energy into acid-base energy, belongs to both the energized oxidized-form type and the energized basic-form type (Fig. 2).

Acknowledgements

This work was supported by grant no. BT45/85 from Comisión Asesora de Investigación Científica y Técnica (Spain). The authors wish to thank Mrs. A. Friend and Mrs. M.J. Pérez de León for helpful secretarial assistance.

REFERENCES

Arnon, D.I. 1977. Photosynthesis 1950-75: Changing concepts and perspectives. In "Encyclopedia of Plant Physiology, New Series, vol 5" (Ed. A. Trebst and M. Avron). (Springer Verlag, Berlin). pp. 7-56.

Babcock, G.T., Widger, W.R., Cramer, W.A., Oertling, W.A. and Metz, J.G. 1985. Axial ligands of chloroplast cytochrome b-559: Identification and requirement for a heme-cross--linked polypeptide structure. Biochemistry, 24, 3638--3645.

Bergström, J. and Vänngard, T. 1982. EPR signals and orientation of cytochromes in the spinach thylakoid membrane. Biochim. Biophys. Acta, 682, 452-456.

Boyer, P.D., Chance, B., Ernster, L., Mitchell, P., Racker, E. and Slater, E. C. 1977. Oxidative phosphorylation and photophosphorylation. Ann. Rev. Biochem., 46, 955-1026.

Galván, F, de la Rosa, F.F., Hervás, M. and Losada, M. 1983. pH-dependent interconversion between the two redox forms of chloroplast cytochrome b-559. Bioelectrochem. Bioenerg., 10, 413-426.

Ghanotakis, D.F., Yocum, C.F. and Babcock, G.T. 1986. ESR spectroscopy demonstrates that cytochrome b-559 remains low potential in Ca^{2+}-reactivated, salt-washed PS II particles. Photosynthesis Research, 9, 125-134.

Hervás, M., Ortega, J.M., de la Rosa, M.A., de la Rosa, F.F. and Losada, M. 1985. Location and function of cytochrome b-559 in the chloroplast non-cyclic electron transport chain. Physiol. Vég., 23, 593-604.

Hill, R. 1965. The biochemists' green mansions: The photosynthetic electron-transport chain in plants. In "Essays in Biochemistry, vol 1" (Ed. P.N. Campbell and G.D. Greville). (Academic Press, New York). pp. 121-151.

Losada, M. 1978. Energy-transducing redox systems and the mechanism of oxidative phosphorylation. Bioelectrochem. Bioenerg., 5, 296-310.

Losada, M. 1986. A unified concept of energy transduction by biochemical systems. Arch. Biol. Med. Exp., 19, 29-56.

Losada, M. and Guerrero, M.G. 1979. The photosynthetic reduction of nitrate and its regulation. In "Photosynthesis in Relation to Model Systems" (Ed. J. Barber). (Elsevier/North Holland Biomedical Press, Amsterdam). pp.365-408.

Losada, M., Hervás, M., de la Rosa, M.A. and de la Rosa, F.F. 1983. Energy transduction by bioelectrochemical systems. Bioelectrochem. Bioenerg., 6, 205-225.

Losada, M, Hervás, M. and Ortega, J.M. 1987. Photosynthetic assimilation of the primordial bioelements. In "Inorganic Nitrogen Metabolism" (Ed. W.R. Ullrich, P.J. Aparicio, P.J. Syrett and F. Castillo). (Springer-Verlag, Berlin). pp. 3-15.

Ortega, J.M., Hervás, M. and Losada, M. 1987. Redox and acid-base characterization of cytochrome b-559 in photosystem II particles. Eur. J. Biochem. (submitted).

Whatley, F.R. and Losada, M. 1964. The photochemical reactions of photosynthesis. In "Photophysiology, vol 1" (Ed. A.C. Giese). (Academic Press, New York). pp. 111-154.

VARIATIONS IN THE CAROTENOID COMPLEMENT OF THE PIGMENT-PROTEIN COMPLEXES OF RHODOSPIRILLUM RUBRUM

R. M. Lozano and J. M. Ramírez

Unidad de Biomembranas
Centro de Investigaciones Biológicas del CSIC
Velázquez 144, 28006 Madrid, Spain

ABSTRACT

Purified light-harvesting antenna complexes of Rhodospirillum rubrum contained a carotenoid mixture, the composition of which was similar to that of the pool of spirilloxanthin and its precursors which was present in intact bacterial cells. Variations in the composition of the pool due to changes in the nutritional state of the cultures were followed by parallel changes in the carotenoid complement of the antenna complexes. The carotenoid heterogeneity of the antenna was not an artefact of solubilization or purification, because a close correspondence was observed in intact cells between the carotenoid absorption spectrum and the excitation spectrum of carotenoid sensitized bacteriochlorophyll fluorescence. In contrast with the lack of uniformity of the antenna carotenoids, the purified reaction center contained only spirilloxanthin and, in exponentially growing cultures, some monodemethylated spirilloxanthin. The convenience of providing a better protection to the reaction center than to the antenna and the possible localization of the reaction center as the main site of singlet oxygen generation might be the reasons why the more unsaturated carotenoids are preferentially incorporated into the reaction center in vivo.

INTRODUCTION

Carotenes and xantophylls are, along with chlorophylls and specific proteins, natural constituents of the intrinsic membrane complexes, antenna and reaction centers, that carry out the initial processes of photosynthesis. The carotenoids perform a dual functional role since besides acting as accesory pigments that absorb light and transfer excitation energy to chlorophylls, they protect cell components from photodestruction by mediating the de-excitation of triplet state chlorophylls and singlet state oxygen (for a recent review see Siefermann-Harms, 1987). This second, protective function seems to be the main reason why carotenoids are present in all natural photosynthetic membranes.

In purified preparations of antenna complexes obtained from several phototrophic purple bacteria, mixtures of metabolically related carotenoids are usually found (Cogdell and Thornber, 1979). The sole reported exception to this rule seems to be a B880 complex of Rhodospirillum rubrum in which only spirilloxanthin was detected, although Duysens (in Cogdell and Thornber, 1979, pp. 77) suggested that its uniform carotenoid composition could be due to the use of old bacterial cultures as the source of the complex. The situation seems to be the opposite for the bacterial reaction center, since purified preparations obtained from several species have been reported to contain one specific carotenoid each (Cogdell et al., 1976; Thornber, 1971; Van der Rest and Gingras, 1974).

We are investigating further the nature of the carotenoids present in the bacterial photosynthetic complexes. In this report we shall present evidence showing that whereas the R. rubrum antenna contains a mixture of carotenoids the composition of which changes with the nutritional state of the culture, only spirilloxanthin and, in minor amounts, monodemethylated spirilloxanthin are found in association with the R. rubrum reaction center. The current progress of our work on the functional bases of the differences in carotenoid composition between the antenna and the reaction center shall also be presented.

MATERIALS AND METHODS

The wild type strain S1 of R. rubrum and the reaction-centerless mutant strain T102 (del Campo et al., 1981) were grown as described previously (Giménez-Gallego et al., 1978). The methods for isolation of membrane vesicles and for solubilization and purification of antenna and reaction center complexes have been also described in previous reports (Giménez-Gallego et al., 1985, 1986). Extraction and purification of carotenoids from cells, isolated membranes and preparations of purified pigment-protein complexes were as follows: at least 10 volumes of acetone were added to the aqueous sample and, after filtration, the extract was shaken together with the same volume of petroleum ether (b.p. 50-70 ^{o}C). The carotenoid-containing upper phase was washed, first 3 times with 3 volumes of distilled water and then with a mixture of methanol/water (95/5, v/v) till the infrared absorption bands of porphyrins became insignificant. Finally the solution was dried under vacuum at room temperature and the residue was dissolved in a appropriate amount of petroleum ether. Individual carotenoids were resolved by thin layer chromatography on silica gel plates, using a mixture of petroleum ether and acetone (9/1, v/v) as the eluent.

Absorption spectra were obtained with a Shimadzu model UV-3000 spectrophotometer in the split beam mode. Fluorescence spectra were scanned with the same instrument supplemented with a fluorescence attachment and a Hamamatsu R473 multiplier. A solution of 3 g·L^{-1} rhodamine B in polyethyleneglycol was used as a photon counter. The digitized spectra were transferred to a Hewlett Packard microcomputer (model 226) for storage and processing.

RESULTS

The carotenoids of intact cells

Coloured carotenoids are synthetized in R. rubrum by a pathway that has rhodopin as a characteristic intermediate and ends in spirilloxanthin (Liaaen-Jensen et al., 1958, 1961; Fig. 1). In exponentially growing cells from phototrophic cultures, the carotenoid pool is a mixture of metabolic intermediates of the biosynthetic pathway and the end product,

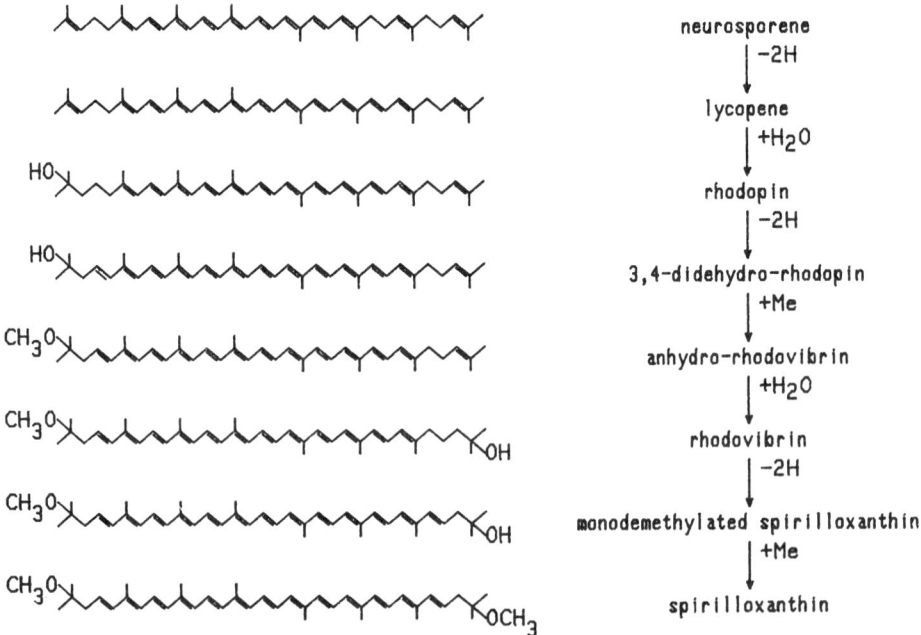

Fig. 1 Biosynthesis of spirilloxanthin in R. rubrum (Liaaen-Jensen et al., 1958, 1961)

spirilloxanthin, amounts only to about a fifth or a tenth of the total carotenoid content. When the cultures use up the carbon source and reach the stationary phase of growth, or when the cells are deprived of carbon source and allowed to stand in the light, the relative content of spirilloxanthin increases gradually until this xantophyll becomes the major constituent of the carotenoid pool (Liaaen-Jensen et al., 1958). Since the visible absorption spectra of carotenoids exhibit a characteristic band which shifts towards lower energies as the number of conjugated double bonds within the pigment molecule increases, the progressive accumulation of the more unsaturated intermediates in cultures which lack source of carbon goes along with a gradual red shift of the composite band of the carotenoid mixture. These changes are illustrated in Fig. 2, where the visible absorption spectra of cells from exponentially-growing and carbon-starved cultures are shown for comparison. The shift of the three-peaked carotenoid band is clearly indicated by the marked variation in the depth of the trough to the left of the $Q_x(0,0)$ bacteriochlorophyll band at 588 nm. There is also a change of the band's fine structure,

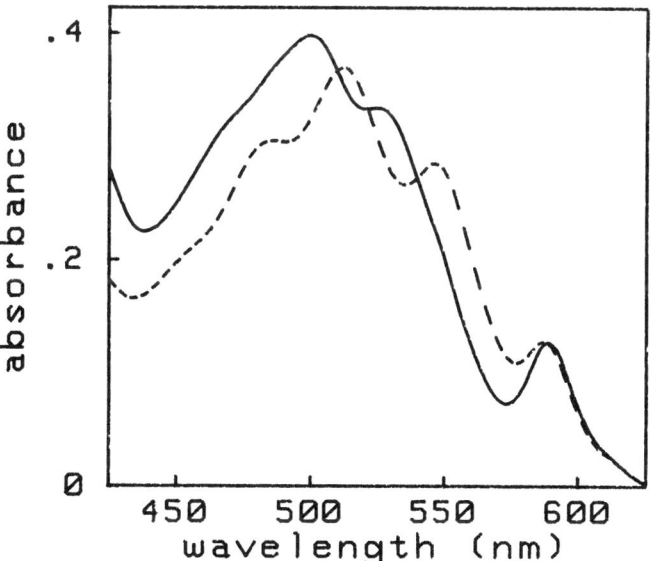

Fig. 2 Evolution of the carotenoid absorption band of R. rubrum cells. Continuous line, cells from midexponential cultures; broken line, carbon starved cells

which is lower in the spectrum of exponential cultures. This additional difference is enhanced by the variation of the extent of overlapping between carotenoid peak III and the $Q_x(0,1)$ band of bacteriochlorophyll near 555 nm.

The carotenoids of isolated pigment protein complexes

In order to investigate whether the carotenoid composition of a culture has any influence on the nature of the pigments which are inserted in the photosynthetic complexes, these were solubilized and purified from several cultures, and their visible absorption spectra were compared with those of the corresponding native membranes. The results are summarized in Table 1, which shows the location of the carotenoid central peak in the spectra of some of the preparations.

The carotenoid absorption band of each purified preparation of antenna complexes was similar, in shape and location, to that of the membranes from which that particular preparation was derived. Besides, the similitude was kept unchanged after organic solvent extraction of the preparations (not shown). Then, it seems that the incorporation of carotenoid pigments to the antenna complexes in vivo is a nonspecific process. The detection of spirilloxanthin as the sole carotenoid of previously analyzed antenna preparations of the same microorganism seems to be accidental and, as suggested by Duysens (in Cogdell and Thornber, 1979, pp. 77), probably due to the use of old, spirilloxanthin enriched cultures as the starting material for complex solubilization.

TABLE 1 Location of the carotenoid band in the spectra of membrane vesicles and purified pigment-protein complexes of R. rubrum

	Wavelength, nm		
	Membrane vesicles	Purified antenna	Reaction centers
In situ			
Exponential cultures	505±1	506±1	503±1
Stationary cultures	511±1	511±1	504±1
In petroleum ether			
Exponential cultures	484±1	-	489±1
Stationary cultures	487±1	-	490±1

In contrast with the purified antenna complexes, reaction centers derived from the same membrane preparations exhibited carotenoid absorption bands that were very similar in both exponential and stationary--phase cultures (Table 1). Organic solvent extraction and purification of the carotenoids present in the reaction centers, followed by thin layer chromatography, indicated spirilloxanthin as the major component. Only its inmediate precursor, monodemethylated spirilloxanthin (Fig. 1) was also present at significant levels in the preparation obtained from exponentially-growing cells. Therefore, the carotenoid binding specificity of the reaction center seems to be as high in R. rubrum as in other phototrophic prokaryotes (Cogdell et al., 1976; Thornber, 1971) and, at any rate, significantly higher than that of the antenna.

Energy transfer from carotenoids to bacteriochlorophyll in the antenna of intact cells

The scarce carotenoid selectivity exhibited by the R. rubrum antenna is striking if one takes into account the usual specificity of protein--ligand interactions. Thus, and in order to exclude possible purification artefacts such as carotenoid exchange between the free and the bound pools during detergent induced solubilization of the complex, the nature of the carotenoid constituents of the antenna was investigated by a different and complementary approach. In the photosynthetic membrane, efficient singlet-singlet energy transfer from excited carotenoid to ground state chlorophyll requires that both pigments are constituents of the photosynthetic complexes. In consequence, the analysis of the excitation spectrum of antenna chlorophyll fluorescence may yield direct information on the absorption spectra of the carotenoids which, being bound to the complexes, are able to sensitize chlorophyll fluorescence. In oxygenic phototrophs, the feasibility of this type of studies is limited by the overlapping of the soret bands of chlorophylls and the visible bands of carotenoids in the violet-blue range of the spectrum. Luckily, this does not apply to R. rubrum and other bacteriochlorophyll containing prokaryotes because the soret bands of these porphyrins are located at higher energies and each type of pigment can be excited separately.

Table 2 shows the location of the carotenoid band in the spectra (absorption and excitation of bacteriochlorophyll fluorescence) of R.

TABLE 2 Correspondence between the carotenoid bands in the absorption and the bacteriochlorophyll-fluorescence excitation spectra of R. rubrum cells

	Wavelength, nm	
	Absorption	Fluorescence
Wild type strain S1		
Exponential cultures	500+1	500±2
Carbon starved cells	512±1	513±3
Reaction-centerless mutant T102		
Exponential cultures	503±1	503±2
Stationary cultures	513±1	514±1

rubrum intact cells which had a variable carotenoid content due to differences in the phase of growth of the corresponding cultures. In addition to phototrophic cultures of wild-type R. rubrum (strain S1), dark, oxygen limited cultures of nonphototrophic strain T102 were also used because in this (as in all other) reaction centerless mutant, antenna bacteriochlorophyll exhibits a significantly increased quantum yield of fluorescence (Heathcote and Clayton, 1977), a property that facilitates the measurement of the excitation spectra in diluted suspensions of intact cells. In all cultures tested the location of the carotenoid band coincided for both the absorption and the excitation spectra. Thus, these results indicate that the carotenoid mixture which is bound to the antenna complexes has a composition identical, or very similar, to that of the total carotenoid pool in the cell, in accordance with the conclusion drawn above from the data obtained with solubilized antenna preparations (Table 1).

DISCUSSION

The results of this study demonstrate that the antenna and the reaction center of R. rubrum differ in their specificities of carotenoid binding. Thus, the microorganism follows in this respect the pattern of other related phototrophic bacteria (Cogdell and Thornber, 1979; Cogdell et al., 1976). Such difference suggests strongly that the functional role of the carotenoid in each type of photosynthetic complex has differential aspects of importance sufficient to impose distinctive structural requirements. The available information on the mechanisms of carotenoid protection in photosynthetic tissues allows to formulate a working hypothesis on the possible nature of such differential aspects.

Aerobic photodestruction of the photosynthetic apparatus seems to be caused mainly by singlet oxygen, which is generated by energy transfer from occasionally formed excited triplet state chlorophyll to ground triplet state oxygen (Siefermann-Harms, 1987),

$$^3Chl^* + {}^3O_2 \rightarrow {}^1Chl + {}^1O_2^* \qquad (1)$$

Carotenoids perform their protective function by quenching both triplet state chlorophyll and single state oxygen:

$$^1Car + {}^3Chl^* \rightarrow {}^3Car^* + {}^1Chl \qquad (2)$$
$$^1Car + {}^1O_2^* \rightarrow {}^3Car^* + {}^3O_2 \qquad (3)$$

Then, the carotenoid triplet state decays via non-radiative intersystem crossing:

$$^3Car^* \rightarrow {}^1Car \qquad (4)$$

While pathway (2) requires carotenoids with 7 or more conjugated double bonds, only carotenoids with 9 or more conjugated double bonds may quench singlet oxygen through pathway (3). It has been estimated that about 90 % of the triplet state chlorophyll decays by pathway (2), and that pathway (3) appears to be only of minor significance in photosynthetic membranes.

At least two explanations may be offered to account for the higher carotenoid binding specificity of the reaction center. The first one is based on the assumption that, as a natural sink for excitation energy, reaction center chlorophyll has a higher probability of populating the excited triplet state and, therefore, of being the sensitizer of singlet oxygen generation through pathway (1). Then, the reaction center would be the more appropriate site for pathway (2) which, involving two nondiffusible molecules, would be favoured by proximity and precise orientation. Such definite spatial requirements would imply a certain degree of carotenoid binding specificity in the complex.

Alternatively -or additionally- the preference of the more unsaturated carotenoids for the reaction center could arise from the convenience of providing the best protection available to this complex, far more expensive in metabolic terms than those of the antenna.

These assumptions may be experimentally tested. Thus, our preliminary results show that the degree of protection afforded by the carotenoids to the reaction center exceeds that given to the antenna by a factor of 2 to 3, suggesting that the higher specificity may be directly related to better protection efficiency.

REFERENCES

del Campo, F.F., Gómez, I., Picorel, R. and Ramírez, J.M. 1981. Electrochromic absorption changes induced by pyrophosphate in chromatophores of photoreaction-centerless mutants of Rr. In "Photosynthesis I. Photophysical Processes-Membrane Energization". (Akoyunoglou, ed.) (Balaban International Science Services, Philadelphia). pp. 515-523.

Cogdell, R.J. and Thornber, J.P. 1979. The preparation and characterization of different types of light-harvesting pigment-protein complexes. In "Chlorophyll organization and Energy Transfer in Photosynthesis". (CIBA Foundation Symposium 61, new series). (Excerpta Medica, Amsterdam). pp. 61-79.

Cogdell, R.J., Parson, W.W. and Kerr, M.A. 1976. The type, amount, location and energy transfer properties of the carotenoid in reaction centers from Rhodopseudomonas sphaeroides. Biochim. Biophys. Acta, 430, 83-93.

Giménez-Gallego, G., del Valle-Tascón, S. and Ramírez, J.M. 1978. Photooxidase system of Rhodospirillum rubrum II. Its role in the regulation of cyclic photophosphorylation. Z. Pflanzenphysiol., 87, 25-36.

Giménez-Gallego, G. Rivas, L. and Ramírez, J.M. 1985. Macromolecular contaminants in preparations of native and proteolyzed Rhodospirillum rubrum reaction centers. Physiol. Vég., 23, 571-581.

Giménez-Gallego, G., Fenoll, C. and Ramírez, J.M. 1986. Purification of a light-harvesting B880 complex from wild-type Rhodospirillum rubrum. Anal. Biochem., 152, 29-34.

Heathcote, P. and Clayton, R.K. 1977. Reconstituted energy transfer from antenna pigment-protein to reaction centers isolated from Rhodopseudomonas sphaeroides. Biochim. Biophys. Acta, 459, 506-515.

Liaaen-Jensen, S., Cohen-Bazire, G., Nakayama, T.O.M. and Stanier, R.Y. 1958. The path of carotenoid synthesis in a photosynthetic bacterium. Biochim. Biophys. Acta, 29, 477-498.

Liaaen-Jensen, S., Cohen-Bazire, G. and Stanier, R.Y. 1961. Biosynthesis of carotenoids in purple bacteria: A reevaluation based on considerations of chemical structure. Nature (London), 192, 1168-1172.

Siefermann-Harms, D. 1987. The light-harvesting and protective functions of carotenoids in photosynthetic membranes. Physiol. Plantarum, 69, 561-568.

Thornber, J.P. 1971. The photochemical reaction center of Rhodopseudomonas viridis. Methods Enzymol. 23, 688-691.

Van der Rest, M. and Gingras, G. 1974. The pigment complement of the photosynthetic reaction center isolated from Rhodospirillum rubrum. J. Biol. Chem., 249, 6446-6453.

TOWARDS AN ANALOGUE OF THE BACTERIAL PHOTOSYNTHETIC REACTION CENTRE :
SYNTHESIS OF AN OBLIQUE BIS-PORPHYRIN SYSTEM CONTAINING
A 1,10-PHENANTHROLINE SPACER

S. Noblat, C. Dietrich-Buchecker, J.P. Sauvage

Laboratoire de Chimie Organo-Minérale, UA 422, Institut de Chimie
1, rue Blaise Pascal, F-67000 Strasbourg-Cedex, France

ABSTRACT

　　The synthesis of a molecule consisting of two porphyrin rings (free base or Zn complex) rigidly held by a 1,10-phenanthroline spacer is described. This new molecule in which distance and orientation of the two porphyrins are well-defined may be an interesting model for the understanding of photodriven electron transfer processes within a redox chain. In particular, oblique disposition and large centre-to-centre separation between the two porphyrins, will they favour the slowing down of recombination reactions?

INTRODUCTION

　　Understanding the factors which govern electron transfer processes in natural photosynthesis might be greatly improved by studying synthetic systems that mimic parts of the biological structures. The recent crystallographic analysis of the photosynthetic reaction centre of the bacteria Rhodopseudomonas viridis (Deisenhofer et al., 1984) shows the intriguing mutual arrangement of the tetrapyrrolic rings : except for the special pair of bacteriochlorophylls, each tetrapyrrole forms an interplane angle of 70° ($BChl_1$, $BChl_2$) or 64° ($BChl_2$, BPh) with its closest neighbour, the centre-to-centre separation being 13 Å or 11 Å (Figure 1). It might be conjectured that this special arrangement is of particular importance with respect to the charge separation process.

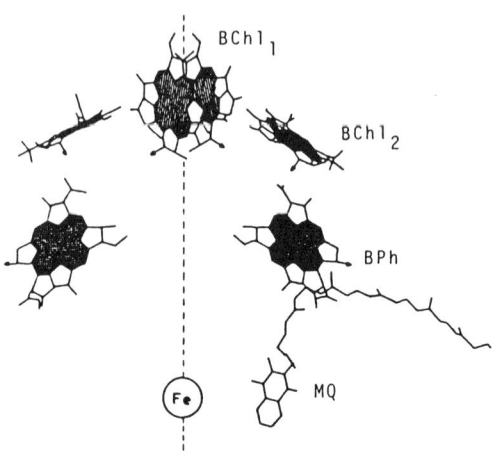

Fig. 1   Molecular arrangement of pigments in bacterial photosynthetic reaction centre from Rhodopseudomonas viridis (BChl : bacteriochlorophyll ; BPh : bacteriopheophytin) (taken from Deisenhofer et al., 1985).

In order to build synthetic analogues of the bacterial reaction centre, several approaches using multi-porphyrin systems have recently been reported (Dubowchik and Hamilton, 1985, 1986 ; Wasielewski et al., 1982). However, the number of molecular arrays whose rigidity is sufficient to ensure control of the distance and orientation of the respective tetrapyrrolic subunits is highly limited (Dubowchik and Hamilton, 1987 ; Cowan et al., 1987, Sessler and Johnson, 1987).

We now report the synthesis of a novel type of "gable" like diporphyrin (Tabushi et al., 1982, 1985), as represented in Figure 2. Compounds 1-3 display geometrical features reminiscent of those found in the reaction centre of R. viridis and more precisely may be an adequate model for the central chromophores $BChl_2$-BPh (centre-to-centre separation between $BChl_2$ and BPh : 11 Å, interplane angle : 64°). In fact, owing to the prefered planar geometry of the 2,9-diphenyl 1,10-phenanthroline fragment (Cesario et al., 1985), both tetrapyrrole units are likely to adopt an oblique disposition. The rotation axis of the two porphyrins should form an angle of 60° and their centre-to-centre separation is about 13 Å (CPK models).

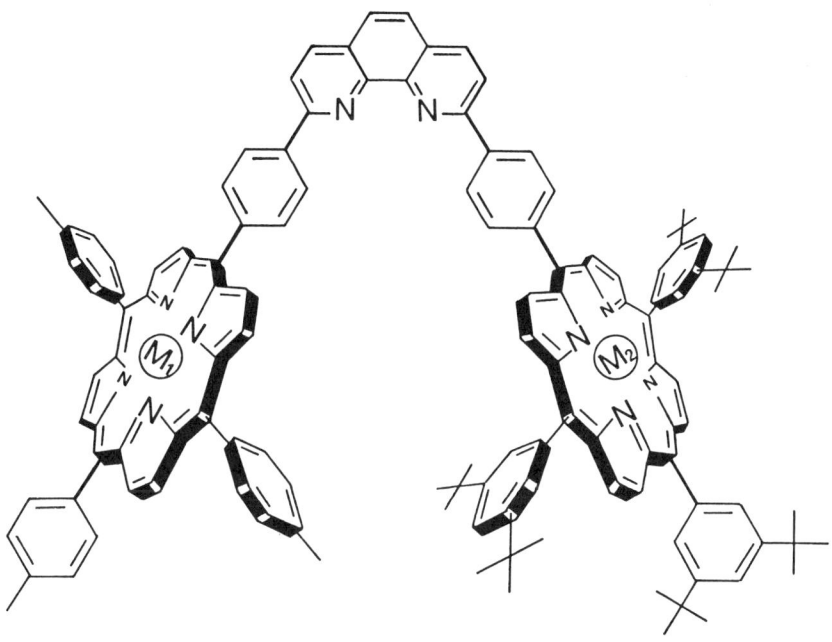

1 : $M_1 = M_2 = 2H$
2 : $M_1 = Zn$, $M_2 = 2H$
3 : $M_1 = M_2 = Zn$

Fig. 2

RESULTS

Compounds 1-3 have been made from 1,10-phenanthroline following a synthetic strategy close to that previously used by Tabushi et al. (1982, 1985) for making a gable-porphyrin. The synthetical pathway and precursors* are represented in Figure 3. 1,10-phenanthroline is reacted with 4-lithiotoluene in excess. After hydrolysis and MnO_2 oxidation following a previously published procedure (Dietrich-Buchecker et al., 1982), 4 is obtained as a white solid (mp 197-199) in 65% yield. Bromination of 4 to 5 by N-bromo succinimid (3 equivalents ; refluxing benzene) is performed under light irradiation. The crude product containing 5 is directly treated by NaOH in refluxing propionic acid.

* : all new compounds gave good elemental analysis. ^1H NMR, UV visible and mass spectra are also consistent with their structures.

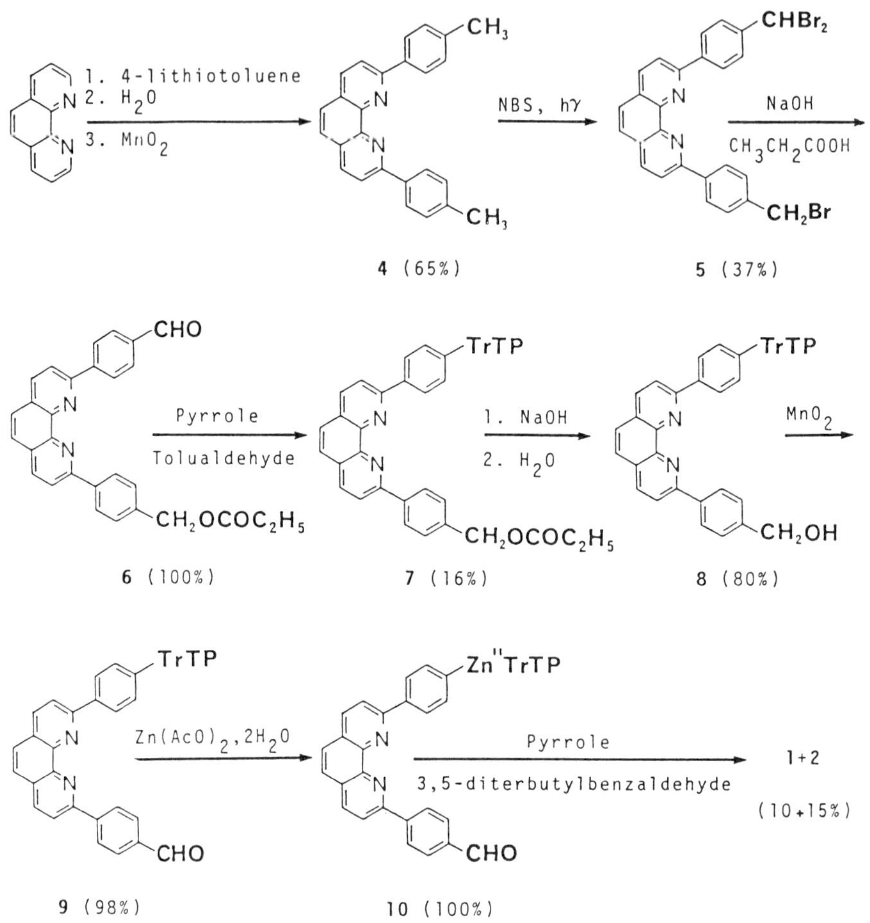

Fig. 3

After work-up, 6 is isolated as a white solid (mp 38-40) in 37% yield as compared to 4. Adlers reaction (Adler et al., 1967) applied to a 1:12:13 mixture of 6, p-tolualdehyde and pyrrole leads to a 16% yield of the monoporphyrin 7, [λ_{max} (log ϵ) (CH_2Cl_2) : 418 (5.29), 517 (4.16), 552 (3.95), 592 (3.64), 648 nm (3.72) ; $M^{\ddagger}$ 996 ; mp > 300], in addition to meso tetra-p-tolyl-porphyrin.

Hydrolysis of 7 with aqueous NaOH furnishes the alcohol 8 in 80% yield. The latter is efficiently oxidized by a large excess of activated MnO_2 : aldehyde 9 is thus obtained as a violet solid [98% ; ^1H NMR shows characteristic peaks at δ 10.10 ppm (CHO) and -2.71 ppm (porphyrinic NH) ; λ_{max} (CH_2Cl_2) : 420, 518, 554, 592, 647 nm].

Quantitative conversion of 9 to its Zinc(II) complex 10 is performed by an excess of $Zn(OAc)_2 \cdot 2H_2O$. The role of the zinc atom within the tri-p-tolyl porphyrin (TrTP) subunit of 10 is to protect the corresponding coordination site and further should allow preparation of hetero-dinuclear complexes. When 3,5 diterbutyl-benzaldehyde (Newman and Lee, 1972) is reacted with aldehyde 10 and pyrrole in boiling propionic acid, a mixture of several porphyrinic compounds is obtained. After repeated chromatographic separations a surprisingly high yield of bis-porphyrins is isolated : 15% of 1 and 10% of 2. In addition are obtained 17% of meso tetra (3,5-di-t-butylphenyl) porphyrin and 9% of its zinc complex. Structural assignement of 1 is based on mass spectroscopy (FAB, $M+H^+$ 1784.8, $C_{127}H_{118}N_{10}$ requires 1784.4), ^1H NMR and electronic spectra [λ_{max} (log ε)(CH_2Cl_2) : 419 (5.60), 517 (4.26), 553 (4.05), 592 (3.75), 648 nm (3.79) ; mp 113-115]. The chemical structure of 2 is demonstrated by the presence of pattern characteristic of both a Zn(II) porphyrin and a free tetrapyrrolic cycle in the visible spectrum [λ_{max} (log ε) (CH_2Cl_2) : 420 (5.57), 517 (4.00), 551 (4.15), 591 (3.68), 648 nm (3.40) ; $M+H^+$: 1848.6, $C_{127}H_{116}N_{10}Zn$ requires 1847.7].

Reaction of 1 or 2 with an excess of $Zn(OAc)_2 \cdot 2H_2O$ leads to quantitative formation of 3, as shown by ^1H NMR, mass and UV-visible spectroscopy [3 : λ_{max} (log ε) (CH_2Cl_2) : 420 (5.29), 550 (4.00), 589 nm (3.36) ; $M+H^+$: 1912.1, $C_{127}H_{114}N_{10}Zn_2$ requires 1911.1].

Preliminary emission measurements show that some type of interaction exists between the two porphyrinic subunits of 3. Indeed, the fluorescence quantum yield of 3 in CH_2Cl_2 (λ_{ex} = 550 nm, λ_{em} = 601 nm) is slightly smaller than that of each constitutive subunit [tetra-p-tolyl-porphyrin or tetra(3,5-di-t-butylphenyl)porphyrin], measured under similar conditions.

REFERENCES

Adler, A.D., Longo, F.R., Finarelli, J.D., Goldmacher, J., Assour, J. and Korsakoff, L. 1967. A simplified synthesis for meso-tetraphenyl-porphin, J. Org. Chem., 32, 476.

Cesario, M., Dietrich-Buchecker, C.O., Guilhem, J., Pascard, C. and Sauvage, J.P. 1985. Molecular structure of a catenand and its copper(I) catenate : complete rearrangement of the interlocked macrocyclic ligands by complexation, J. Chem. Soc. Chem. Commun., 244-247.

Cowan, J.A., Sanders, J.K.M., Beddard, G.S. and Harrison, R.J. 1987. Modelling the photosynthetic reaction centre : photoinduced electron transfer in a pyromellitimide-bridged "special pair" Porphyrin Dimer, J. Chem. Soc., Chem. Commun., 55-58.

Deisenhofer, J., Epp, O., Miki, K., Huber, R. and Michel, H. 1984. X-ray structure analysis of a membrane protein complex, J. Mol. Biol., 180, 385-398.

Deisenhofer, J., Michel, H., Zinth, W., Knapp, E.W., Fischer, S.F. and Kaiser, W. 1985. Correlation of structural and spectroscopic properties of a photosynthetic reaction center, Chem. Phys. Letters, 119, 1-4.

Dietrich-Buchecker, C.O., Marnot, P.A. and Sauvage, J.P. 1982. Direct synthesis of disubstituted aromatic polyimine chelates, Tetrahedron Lett., 23, 5291-5294.

Dubowchik, G. and Hamilton, A.D. 1985. Controlled conformational changes in covalently-linked dimeric porphyrins, J. Chem. Soc., Chem Commun., 904-906.

Dubowchik, G. and Hamilton, A.D. 1986. Towards a synthetic model of the structure of the photosynthetic reaction centre, J. Chem. Soc., Chem. Commun., 1391-1394.

Dubowchik, G. and Hamilton, A.D. 1987. Synthesis of tetrameric and hexameric cyclo-porphyrins, J. Chem. Soc. Chem. Commun., 293-295.

Newman, M.S. and Lee, L.F. 1972. The synthesis of arylacetylenes. 3,5-Di-tert-butylphenylacetylene, J. Org. Chem., 37 4468-4469. This aldehyde was used in order to increase the solubility of the final product.

Sessler, J.L. and Johnson, M.R. 1987. The synthesis of 1,3- and 1,4-phenylene-linked bisquinone-substituted porphyrin dimers, Angew. Chem., 99, 678-680.

Tabushi, I and Sasaki, T. 1982. Gable-porphyrins as a cytochrome-C_3 model, Tetrahedron Lett., 1913-1916.

Tabushi, I. Kugimiya, S. and Sasaki, T., 1985. Artificial allosteric systems. 3. Cooperative carbon monoxide binding to Diiron(II)-Gable Porphyrin-Diimidazolylmethane complexes, J. Am. Chem. Soc., 107, 5159-5163.

Wasielewski, M.R., Niemczyk, M.P. and Svec, W.A., 1982. Selectively metalated doubly cofacial porphyrin trimers. New models for the study of photoinduced intramolecular electron transfer, Tetrahedron Lett., 3215-3218.

LOCATION AND ORGANISATION OF THE CHLOROPHYLL-PROTEINS OF PHOTOSYNTHETIC REACTION CENTRES IN HIGHER PLANTS

D.J. Simpson[1], R. Bassi[2], O. Vallon[1,3], and G. Høyer-Hansen[1]

[1] Department of Physiology, Carlsberg Laboratory,
Gamle Carlsberg Vej 10, DK-2500 Copenhagen, Denmark.
[2] Dipartimento di Biologia, Universita` di Padova,
Via Loredan 10, 35131 Padova, Italy.
[3] Present address: Department of Cellular and Developmental Biology, The Biological Laboratories, Harvard University, Cambridge MA 02138, USA.

ABSTRACT

The thylakoids of higher plants are organised into appressed and non-appressed regions, which can be distinguished by their freeze-fracture ultrastructure. Mutants have been used to localise indirectly, the reaction centres of photosystem I (PSI) and photosystem II (PSII). Thus, PSII is located in the EFs particles in appressed membranes, while PSI is found in the large PFu particles in non-appressed membranes. The extreme lateral heterogeneity of thylakoids (i.e., all PSI in non-appressed membranes) was confirmed by immunogold labelling of isolated thylakoids using monoclonal antibodies raised against the PSI reaction centre polypeptide. Antibodies against the core polypeptides of the PSII reaction centre mostly labelled appressed membranes, although 15-20% of the label was over non-appressed regions. This may be due to non-functional PSII centres in the stroma lamellae (PSIIβ centres). Consistent with this was the exclusive labelling of appressed membranes by a monoclonal antibody to CP29, an antennae of PSII.

PSI particles with a chlorophyll a/b ratio of 6.0 and a chlorophyll:P700 ratio of 208:1 were isolated and designated PSI-200. They could be resolved into 3 different chlorophyll-proteins by non-denaturing gel electrophoresis, or by detergent fractionation and sucrose gradient ultracentrifugation. Analysis of the different fractions led to a model for the organisation of PSI in barley. By slightly reducing the amount of detergent used in solubilising PSI, another preparation, with a chlorophyll a/b ratio of 3.5 and a chlorophyll:P700 ratio of 300:1 was isolated in which LHCII was present and functionally connected to the reaction centre. Reconstitution experiments showed that LHCI-680 was necessary for energy transfer to PSI from LHCII. The following excitation energy transfer sequence is proposed:

$$LHCII \rightarrow LHCI\text{-}680 \rightarrow LHCI\text{-}730 \rightarrow P700\ Chl_a\text{-}P1$$

Purified PSII particles were prepared from isolated appressed membranes and analysed by an improved gel electrophoretic system which maintained all the LHCII in the oligomeric form. This revealed 3 chlorophyll a/b-proteins in the 27 kDa region, designated CP29, CP26 and CP24. A new chlorophyll-protein complex, designated CP47*, was also found, and this was capable of electron transport from diphenylcarbazide to 2,6-dichlorophenolindophenol. Detergent solubilisation in the presence or absence of magnesium ions, followed by sucrose gradient ultracentrifugation, provided information about the organisation of the different chlorophyll-proteins of PSII. Most of the excitation energy is postulated to be transferred to the reaction centre in the following sequence:

$$LHCII \rightarrow CP43 \rightarrow CP47 \rightarrow P680$$

and the rest comes to CP43 from CP29 and CP26 via CP24. A tentative model for the organisation of PSII is presented.

INTRODUCTION

The conversion of solar energy into high energy chemical compounds occurs in the chloroplasts of plants and algae. Each plant cell contains about 50 of these organelles which contain their own DNA and protein-synthesising machinery, and divide semi-autonomously. Each chloroplast is surrounded by a double membrane envelope and contains a soluble protein component known as the stroma, where the carbon-fixing Calvin cycle enzymes are located, and an extensive folded membrane system called the thylakoid membrane, which is the site of energy trapping and its conversion into ATP and NADPH (Figure 1).

Our investigations into the structure and organisation of the chloroplast thylakoid membrane have involved the following approaches:

1. isolation of single nuclear gene mutants in barley affecting photosynthesis or pigment synthesis, and their genetic characterisation
2. characterisation of these mutants by spectroscopy, electron transport reactions, polypeptide composition and chlorophyll-protein content
3. structural examination by thin section and freeze-fracture electron microscopy
4. isolation of polypeptide components, raising of monoclonal antibodies and using them to detect specific proteins in mutants by immunoblotting and to determine the location of polypeptides in thylakoids by immunocytochemistry with immunogold labelling
5. isolation of purified, functional reaction centre preparations, followed by detergent fractionation and characterisation.

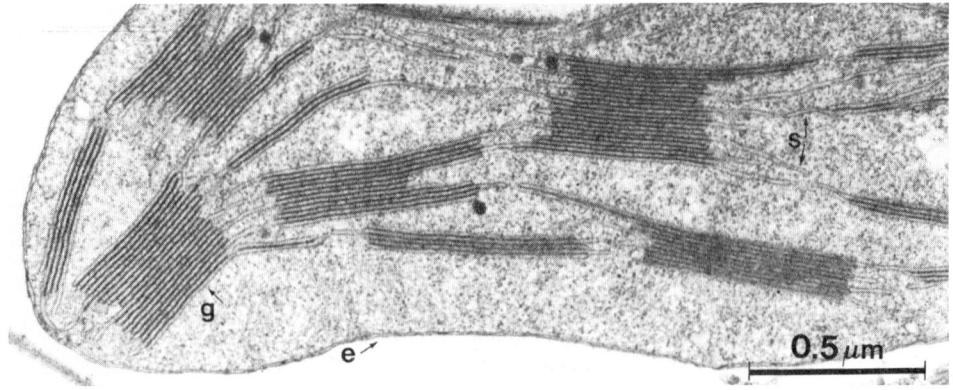

FIGURE 1. Thin section of part of an isolated chloroplast showing the internal thylakoid membrane system which consists of appressed grana lamellae (g) and non-appressed stroma lamellae (s) embedded in the stroma protein matrix and surrounded by a double membrane envelope (e).

The present paper summarises our recent results relating to the location and structure of the reaction centres of photosystems I and II, with particular reference to the organisation of the chlorophyll-proteins and the transfer of excitation energy from antennae to the reaction centres.

ULTRASTRUCTURAL INVESTIGATIONS

Photosystem I

Mechanical disruption of thylakoids into appressed (grana lamellae) and non-appressed (stroma lamellae) membrane fractions showed that photosystem I was preferentially located in the stroma lamellae (Sane et al., 1970), and this has been confirmed by thylakoid fractionation and separation by aqueous two-phase partitioning (Andersson and Anderson, 1980). Barley mutants deficient in, or completely lacking photosystem I, have thylakoids which look very similar to wild type plants when examined by thin section electron microscopy (Simpson, 1982, 1983). When isolated thylakoids were examined by freeze-fracture electron microscopy, it became clear that the large particles on the protoplasmic fracture face of stroma lamellae (PFu) of wild type thylakoids (Figure 2a), were reduced in number in the partially deficient mutant viridis--n^{34}, and absent from the mutant viridis-zb^{63} (Figure 2b), which has no photosystem I activity and completely lacks the P700 Chl_a-P1 reaction centre

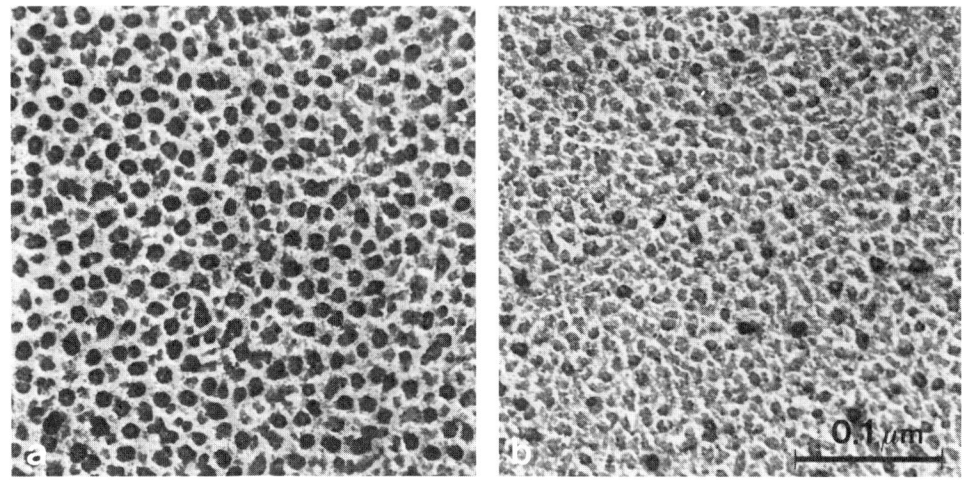

FIGURE 2. High magnification freeze-fracture electron micrograph of the protoplasmic face of non-appressed thylakoids (PFu face). Wild type membranes (a) have both large and small particles, while the photosystem I mutant viridis-zb^{63} (b) contains only small particles, indicating that the photosystem I reaction centre is located in the large particles on the wild type PFu face.

polypeptide (Hiller et al., 1980). No ultrastructural differences were seen between the granal membranes of mutant and wild type thylakoids. It was therefore concluded that the photosystem I reaction centre was located in the large PFu particles revealed by freeze-fracture (Simpson 1983). Analysis of a Chlamydomonas mutant lacking the light-harvesting complex of photosystem I (LHCI or CP0) showed that LHCI was also located in the large PFu particles (Olive et al. 1983).

The exclusive localisation of photosystem I in stroma lamellae was demonstrated directly by immunogold labelling of isolated thylakoids (Vallon et al. 1985, 1986, 1987b). Monoclonal antibodies raised against the barley photosystem I reaction centre (P700 Chl_a-P1 or CPI) are incubated with thin sections of wild type barley thylakoids. The antibody binds specifically to CPI and is visualised by incubation with a rabbit anti-mouse antibody which was labelled with 12 nm colloidal gold (Høyer-Hansen et al., 1987, Vallon et al., 1987a). Gold particles are located exclusively over stroma lamellae and the non-appressed end membranes of grana (Figure 3a). Numerical analysis of the distribution of gold particles showed a 22-fold higher labelling density over non-appressed membranes compared with appressed membranes (Table 1). This represents further evidence for the concept of extreme lateral heterogeneity in the composition of the thylakoid membrane (Andersson and Anderson, 1980).

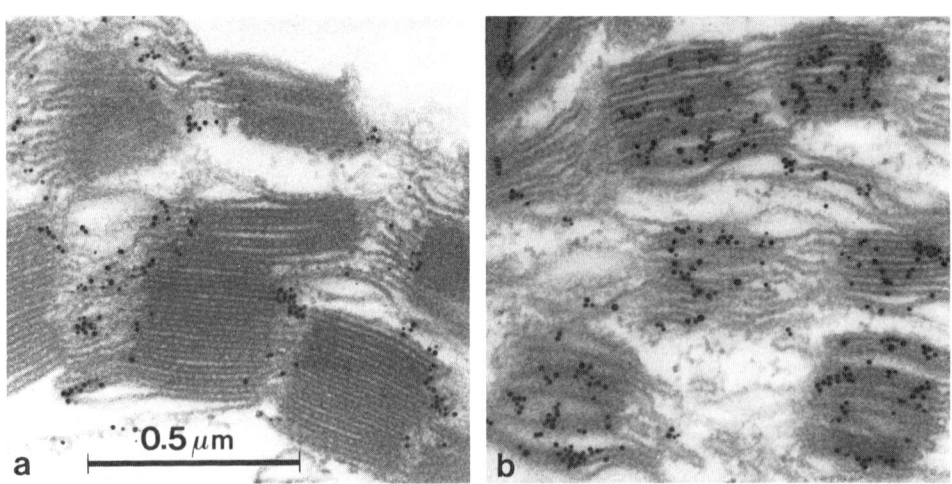

FIGURE 3. Immunogold labelling of thin sections of isolated wild type membranes incubated with monoclonal antibodies raised against (a) the reaction centre polypeptide of PSI (P700 Chl_a-P1) and (b) the PSII antenna chlorophyll-protein CP29. The binding of the antibodies is visualised by a subsequent incubation with colloidal gold-labelled rabbit anti-mouse antibody. PSI is located exclusively in non-appressed lamellae, while CP29 is found mainly over appressed lamellae.

TABLE 1. Immunogold labelling of barley thylakoids.

Antibody to	% label on stacked regions	non-appressed* / appressed
Chl_a-Pl	7.7	22.4
D1	81.0	0.44
CP47	80.5	0.45
CP43	84.3	0.35
CP29	95.4	0.090

* this value gives the relative labelling densities of gold particles on non-appressed versus appressed membranes.

Photosystem II

Photosystem II reaction centres have been shown to be located mainly in the appressed granal membranes by a number of independent but indirect methods, including thylakoid fractionation, de-stacking and re-stacking experiments and chemical modification. Further evidence has come from freeze-fracture investigations of mutants specifically deficient in photosystem II in barley (Simpson et al., 1977), tobacco (Miller and Cushman, 1979) and Chlamydomonas (Olive et al. 1979). In all cases, there was a partial or almost total loss of the particles on the endoplasmic fracture face of grana (EFs), but not stroma lamellae (EFu). Thus the mutant viridis-115 totally lacks photosystem II, but is otherwise indistinguishable from wild type and has almost none of the EFs particles (Figure 4b) found in the wild type (Figure 4a). The presence of EFs particles is positively correlated with tetrameric particles on the inner thylakoid surface (ESs), which have recently been shown to be the site of the extrinsic polypeptides of the oxygen evolving complex of photosystem II (Simpson and Andersson, 1986).

Partial loss of photosystem II activity and EFs particles can be induced by manganese deficiency in barley and spinach (Simpson and Robinson, 1984). In the presence of limiting amounts of manganese, which is an essential component of the oxygen evolving complex, the number of EFs particles is reduced to 25% of the control, with a corresponding loss of photosystem II activity. These plants, which resemble partial photosystem II mutants of barley (Simpson, 1981) and Chlamydomonas (Olive et al., 1979), recover within 2-3 days of manganese application. As with the photosystem II mutants, the number of particles on the EFu face of manganese-deficient leaves is not affected. These may represent the photosystem II reaction centres in stroma lamellae, the so-called PSIIβ centres (Melis, 1986), but there is no direct evidence for this. Between 15-20% of the core polypeptide components of photosystem II are located in the stroma lamellae (Table 1). It is of interest, therefore, that the minor photosystem II antennae protein called CP29 is located exclusively in appressed granal membranes (Figure 3b), and is not, therefore, a component of the photosystem II reaction centres located in the stroma lamellae.

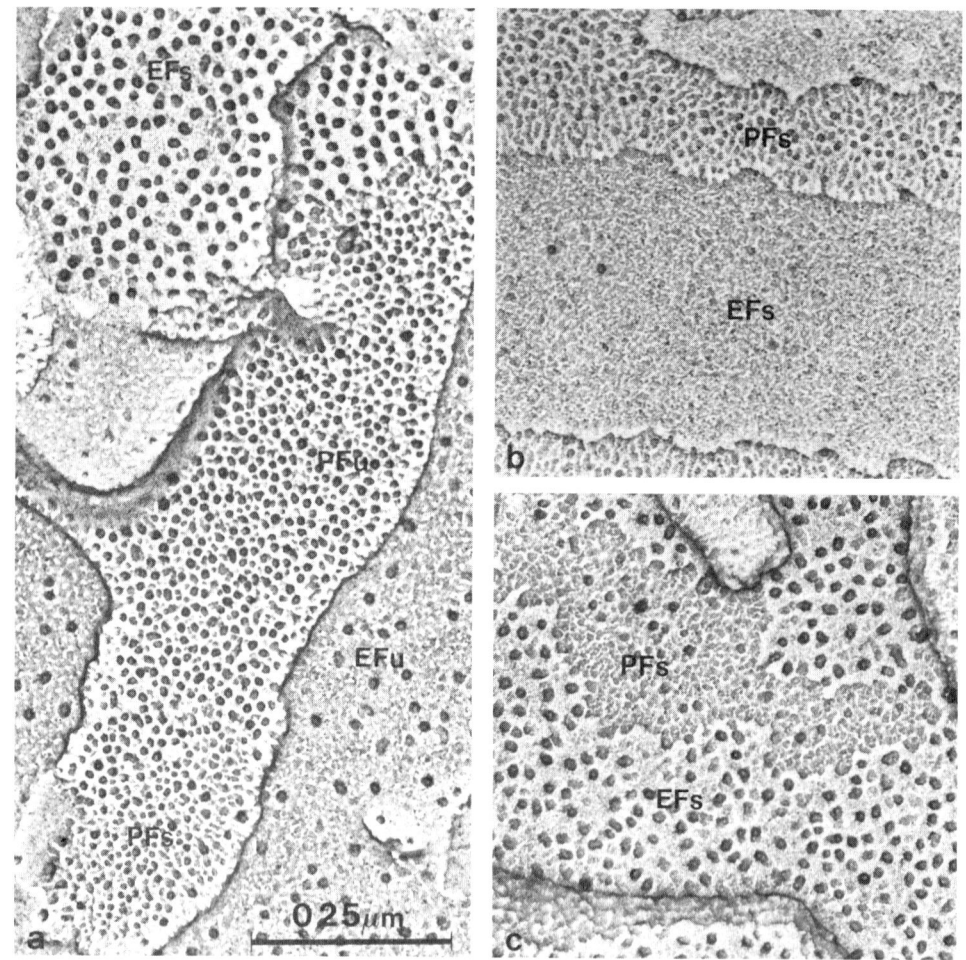

FIGURE 4. Freeze-fracture electron micrographs of thylakoids from (a) wild type, (b) the photosystem II mutant viridis-115 and (c) the chlorophyll b-less mutant chlorina-f2. The EFs particles found in the wild type (a) are almost all missing from the photosystem II mutant (b), indicating that the photosystem II reaction centre is located in these particles. The PFs particles are missing from the chlorophyll b-less mutant (c), showing that LHCII, the major chlorophyll-containing protein is the major constituent of the PFs particles in wild type thylakoids.

In contrast to LHCI, the light-harvesting chlorophyll a/b-antennae complex of photosystem II (LHCII) is the major component of the particles on the complementary protoplasmic fracture face of appressed membranes (PFs) (Simpson, 1979, Olive et al., 1979), and does not appear to be a significant component of the reaction centre EFs particles, although this is disputed. The LHCII in PFs particles is, nevertheless, in contact with the reaction centre particles and may provide a pathway for excitation energy transfer between several photosystem II reaction centres.

DETERGENT FRACTIONATION

Photosystem I

The isolation of purified photosystem I is based on its selective solubilisation by addition of the non-ionic detergent Triton X-100 to isolated, de-stacked thylakoids (Mullet et al., 1980). Thylakoids are suspended to 0.8 mg chlorophyll/ml and Triton X-100 added to a final concentration of 0.6-0.75% and photosystem I is isolated by sucrose gradient ultracentrifugation (Bassi et al., 1985, Bassi and Simpson, 1987a,c). This produces a functional photosystem I preparation which fluorescences at 730 nm, with a chlorophyll a/b ratio of 6.0 and a chlorophyll:P700 ratio of 208:1 (PSI-200). Analysis by non-denaturing gel electrophoresis shows the presence of 3 different chlorophyll-proteins: the reaction centre P700 Chl_a-P1 (CPI) and 2 chlorophyll a/b-containing antennae designated LHCI-730 and LHCI-680. When these chlorophyll-proteins are eluted from the polyacrylamide gel and analysed by spectroscopy, they each have characteristic circular dichroism spectra and low temperature absorption and fluorescence emission spectra. The fluorescence emission maxima are particularly characteristic, being 720 nm for P700 Chl_a-P1, 730 nm for LHCI-730 and 680 nm for LHCI-680 (Table 2). Re-electrophoresis under denaturing conditions in the presence of 6 M-urea has shown that each chlorophyll-protein of photosystem I has its own unique polypeptide composition (Table 2).

The PSI-200 preparation can be subjected to further fractionation by solubilisation in the detergent dodecyl maltoside, followed by sucrose gradient ultracentrifugation (Bassi and Simpson 1987c). This produces up to 6 well-resolved chlorophyll-containing bands, 4 of which are of special interest. One consists of pure LHCI-680, which is identical in its composition to that isolated by gel electrophoresis, but which fluoresces at 77°K with a maximum at 690 nm, probably reflecting milder treatment during isolation. A second band contains pure LHCI-730, which is identical to that isolated from PSI-

TABLE 2.   Properties of thylakoid chlorophyll-proteins.

Name	Other names	Chl a/b ratio	apparent mw (kDa)	fluorescence (nm)	genes	
Photosystem I:						
CPI	Chl_a-P1	∞ =	68, 67	720	psaA,B	(c)[§]
LHCI-680	LHCIa	3.7	24, 20	690		(n)
LHCI-730	LHCIb	3.7	23, 22, 21	730		(n)
Photosystem II:						
CP47	Chl_a-P2, CPa	∞	47	682	psbB	(c)
CP43	Chl_a-P3, CPa	∞	43	682	psbC	(c)
D1	Q_B-binding	∞	32	-	psbA	(c)
D2	Q_A-binding (?)	∞	32-34	-	psbD	(c)
LHCII	$Chl_{a/b}$-P2**, CPII*, CP27	1.3	29.5, 29, 28.5, 28, 26	682, 695	cab	(n)
CP29	$Chl_{a/b}$-P1	2.0	29	680		(n)
CP26	$Chl_{a/b}$-P2, CP27	1.8	28.5, 28	680		(n)
CP24	$Chl_{a/b}$-P3	1.2	24, 20	680		(n)

[+] i.e., chlorophyll a only
[§] c = chloroplast gene, n = nuclear gene

200 by gel electrophoresis. The third band is PSI-200 stripped of LHCI - it contains no chlorophyll b, fluoresces at 720 nm, and has a chlorophyll:P700 ratio of 90:1 (PSI-90). The fourth band contains P700 Chl_a-P1 and varying amounts of LHCI-730, but no LHCI-680, and fluoresces at 730 nm. There is no fluorescence peak at 720 nm from P700 Chl_a-P1, which must be quenched by LHCI-730. Based on the fluorescence emission properties of the different photosystem I preparations, the following sequence of excitation energy transfer is proposed:

LHCI-680 → LHCI-730 → P700 Chl_a-P1

The distribution of chlorophyll a and b among the different fractions separated from PSI-200 by sucrose gradient ultracentrifugation was analysed. This revealed that about one-quarter of the chlorophyll b was associated with LHCI-680, and the rest in LHCI-730. Since there were several different photosystem I preparations with different chlorophyll a/b and chlorophyll:P700 ratios, it was possible to calculate the chlorophyll:P700 ratio in P700 Chl_a-P1 and the average chlorophyll a/b ratio of LHCI (Bassi and Simpson, 1987c). The results have been used to produce the hypothetical model of photosystem I shown in Figure 5.

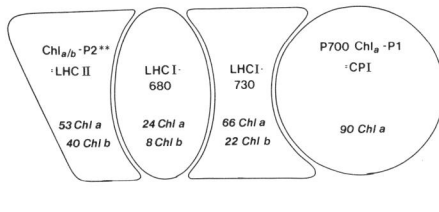

FIGURE 5. Model for the organisation of photosystem I in barley, showing the distribution of chlorophyll a and b between the different chlorophyll-proteins. Over half of the total chlorophyll is in LHCI, and LHCI-680 is postulated to mediate excitation energy transfer from LHCII, when LHCII becomes phosphorylated and migrates from appressed to non-appressed membranes.

By reducing the amount of Triton X-100 relative to the chlorophyll concentration in the initial solubilisation of thylakoids, a denser photosystem I preparation was isolated from the sucrose gradient, with a chlorophyll a/b ratio of 3.5, and a chlorophyll:P700 ratio of 300:1 (PSI-300). Analysis of the chlorophyll-proteins revealed the presence of large amounts of LHCII - the light-harvesting chlorophyll a/b protein of photosystem II. The fluorescence emission spectrum showed a single peak at 731 nm, with none of the 682 nm fluorescence characteristic of isolated LHCII. When LHCII was disconnected from the photosystem I reaction centre by the addition of 0.4% octyl glucoside, the fluorescence yield at 682 nm increased, with a corresponding decrease at 731 nm. This technique was used in reconstitution experiments with isolated LHCII and different photosystem I preparations to show that LHCII became functionally connected to photosystem I only when LHCI-680 was also present. Thus, up to 100 molecules of chlorophyll as LHCII can be functionally associated with the photosystem I antennae in vitro, and although this is an artefact of the isolation procedure, it demonstrates that LHCII can transfer excitation energy to photosystem I, which probably happens in the thylakoid as a result of LHCII phosphorylation after the state 1 - state 2 transition. The proposed excitation energy transfer sequence is thus:

LHCII → LHCI-680 → LHCI-730 → P700 Chl_a-P1

These results are summarised in Figure 6, which incorporates the earlier concepts of Mullet et al., (1980) and Ish-Shalom and Ohad (1983).

Photosystem II

It is not so long ago that the only known chlorophyll-proteins were CPI (P700 Chl_a-P1) and CPII (LHCII). As electrophoretic techniques have improved, and conditions for thylakoid solubilisation become more gentle, an increasing proportion of the chlorophyll remains non-covalently attached to the chlorophyll-proteins resolved by polyacrylamide gel electrophoresis. Since the

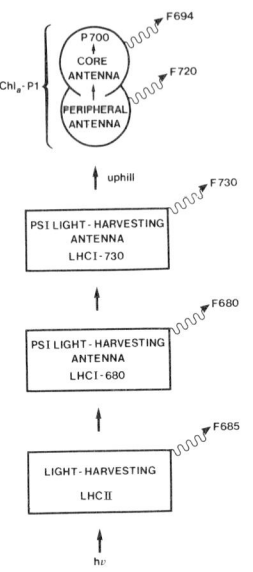

FIGURE 6. Postulated excitation energy transfer sequence between the chlorophyll-proteins of photosystem I, with low temperature fluorescence emission peaks shown. This model has proven useful in predicting and interpreting the fluorescence emission spectra of barley mutants lacking one or more of the chlorophyll-proteins of photosystem I. Thus the viridis-k^{23} mutant fluoresces at 720 nm and completely lacks LHCI, whereas the chlorophyll b-less mutant fluorescences at 730 nm, and LHCI can be detected by immunoblotting.

function of many of the chlorophyll-proteins of photosystem II remains in doubt, most are referred to by their apparent molecular mass. Thus the designation CP47 refers to a chlorophyll-protein with an apparent molecular mass of 47 kDa. Functional photosystem II preparations can be isolated by Triton X-100 treatment of isolated, stacked thylakoids, which causes a selective solubilisation of the non-appressed lamellae, and the photosystem II-rich appressed lamellae can be isolated by centrifugation (Berthold et al., 1981). These can be used as the starting material for the analysis of the structure of the photosystem II reaction centre.

The use of the detergent octyl glucoside to solubilise photosystem II membranes, plus the incorporation of glycerol in the solubilisation buffer and the gel, produces a significant improvement in the separation of photosystem II chlorophyll-proteins (Bassi and Simpson, 1987b, Bassi et al. 1987). All the LHCII, which normally migrates as a monomer with an apparent molecular mass of 27 kDa, is maintained in the oligomeric form, with an apparent molecular mass of 64 kDa. This enables 3 different chlorophyll a/b-proteins to be resolved in the 27 kDa region, and they are designated CP29 (see also Figure 3b), CP26 and CP24 (Table 2). All 3 have different polypeptide compositions, chlorophyll a/b ratios and absorption spectra, but have the same low temperature fluorescence emission maxima (Table 2). None of them become phosphorylated when thylakoids are incubated with $[\gamma-^{32}P]$ATP under conditions which extensively label LHCII. CP29 was first isolated by Machold and Meister

(1979) and CP24, which was first described by Dunahay and Staehelin (1986), appears to be similar, if not identical to LHCI-680 (Bassi et al., 1987).

In addition to these 4 chlorophyll a/b-proteins (LHCII, CP29, CP26, CP24), 3 chlorophyll a-containing bands are resolved by this gel system. Two of them correspond to CP43 and CP47, the latter being present at an unusually low level. The third band has the same mobility as P700 Chl_a-P1, but contains only CP47, plus the proteins D1, D2 and cytochrome b-559, and is designated CP47* (or Chl_a-P2*). It fluoresces at 690 nm and has the simplest composition of any preparation capable of photosystem II dependent electron transport (from diphenyl carbazide to 2,6-dichlorophenolindophenol). It corresponds to the CP2-b preparation of Yamagishi and Satoh (1985), and demonstrates that CP43 does not contain the photosystem II reaction centre. The recent isolation of a photosystem II particle containing only D1, D2 and cytochrome b-559 and capable of photo-oxidising P680, shows that the reaction centre is probably located in D1 and D2 (Namba and Satoh, 1987), analogous to the L and M subunits of the purple bacteria (Rhodopseudomonas viridis) reaction centre (Michel and Deisenhofer, 1986), and not in CP47.

The nearly identical fluorescence emission spectra of the different chlorophyll-proteins of photosystem II (Table 2) makes it impossible to use fluorescence spectroscopy to determine the excitation energy transfer sequence in photosystem II, as was done for photosystem I. Instead, photosystem II membranes are solubilised with octyl glucoside in the presence or absence of magnesium ions, which are known to stabilise the intermolecular binding between monomers of LHCII. Solubilised photosystem II membranes are then separated by sucrose gradient ultracentrifugation, and the resulting bands analysed for chlorophyll-protein and polypeptide composition (Table 3). In the presence of Mg^{++}, 2 of the chlorophyll-containing bands are complementary in their composition, suggesting that each contains a single complex. One consists of CP29, CP26 and CP24, while the other contains LHCII, CP47, CP43 and CP24. Since CP24 is a component of both bands, it may link CP29 and CP26 with the rest of the photosystem II complex.

In the absence of Mg^{++}, 4 bands are produced, 2 of which (bands II and IV) are complementary in their chlorophyll-protein composition (Table 3). Band IV is identical to the CP47* complex isolated by gel electrophoresis, with a slight contamination by CP43. CP43 is a component of all bands, and accounts for the difference in composition of bands I and II. It is proposed that CP43 is important in binding CP47* to the antennae proteins. These results are summarised in Figure 7. The structural relationships between the

TABLE 3. Chlorophyll-protein composition of the green bands separated by sucrose gradient ultracentrifugation of photosystem II.

band	+ Mg^{++}	- Mg^{++}
I	CP29, CP26, CP24	CP43 (minor), CP29, CP26, LHCII**, CP24
II	CP47, CP43, LHCII**, CP24 (trace)	CP43, CP29, CP26, LHCII**, CP24
III	CP47, CP43, LHCII**, CP24	CP47, CP43, CP29, CP26, LHCII**, CP24
IV	-	CP47, CP43(minor)

component chlorophyll-proteins of photosystem II have been used as the basis of the predicted excitation energy transfer pathways in Figure 8. Most of the excitation energy arriving at the reaction centre (in CP47*) comes from LHCII via CP43. There is a significant number of chlorophyll molecules in CP29, which together with CP26 may transfer excitation energy to CP43 via CP24. We do not know how significant these chlorophyll-proteins are as a source of excitation energy arriving at the photosystem II reaction centre. They may be important in diverting excess excitation energy away from closed centres. Under photoinhibitory conditions, this would protect the photosystem II reaction centre by providing a non-destructive means of energy dissipation.

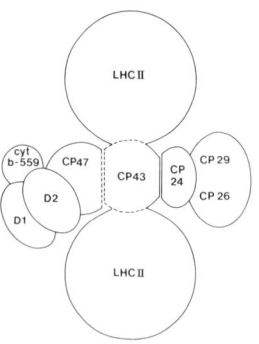

ORGANISATION OF PS II

FIGURE 7. Tentative model for the organisation of photosystem II, showing the relationship between the antennae chlorophyll-proteins and the reaction centre. The dotted lines show bonds stabilised by the presence of Mg^{++}. The diagram is drawn parallel to the plane of the membrane, so that the components extend above and below the plane of the page.

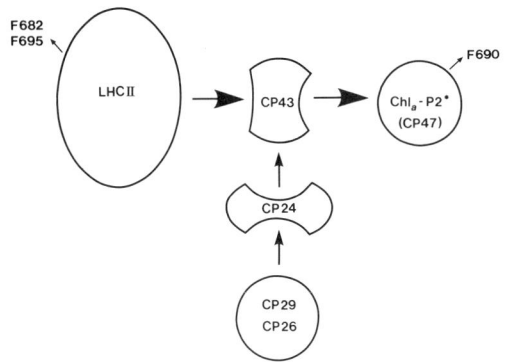

FIGURE 8. Hypothetical scheme for excitation energy transfer within photosystem II, based largely on the structural organisation of the chlorophyll-proteins shown in the model in Figure 7.

ACKNOWLEDGEMENTS

We wish to thank Prof. Diter von Wettstein for his encouragement and helpful discussions during the course of this work. We are grateful to Nina Rasmussen and Ann-Sofi Steinholtz for the drawings and photography.

REFERENCES

Andersson, B. and Anderson J.M. 1980. Lateral heterogeneity in the distribution of chlorophyll-protein complexes of the thylakoid membranes of spinach chloroplasts. Biochim. Biophys. Acta 593, 427-440.

Bassi, R., Machold O. and Simpson D.J. 1985. Chlorophyll-proteins of two photosystem I preparations from maize. Carlsberg Res. Commun. 50, 145-162.

Bassi, R. and Simpson D.J. 1987a. Light-harvesting chlorophyll-proteins of barley photosystem I. In: Prog. Photosynthesis Res. (Ed. J. Biggins), (Martinus Nijhoff Publ., Dordrecht, Netherlands). Vol. II, 61-64.

Bassi, R. and Simpson D.J. 1987b. The organisation of photosystem II chlorophyll-proteins. In: Prog. Photosynthesis Res. (Ed. J. Biggins), (Martinus Nijhoff Publ., Dordrecht, Netherlands). Vol. II, 81-88.

Bassi, R. and Simpson D.J. 1987c. Chlorophyll-protein complexes of barley photosystem I. Eur. J. Biochem. 163, 221-230.

Bassi, R., Høyer-Hansen G., Barbato R., Giacometti G.M. and Simpson D.J. 1987. Chlorophyll-proteins of the photosystem II antennae system. J. Biol. Chem. 262, in press.

Berthold, D.A., Babcock G.T. and Yocum C.F. 1981. A highly resolved, oxygen-evolving photosystem II preparation from spinach thylakoid membranes. EPR and electron transport properties. FEBS Lett. 134, 231-234.

Dunahay, T. and L.A. Staehelin. 1986. Isolation and characterization of a new minor chlorophyll a/b-protein complex (CP24) from spinach. Plant Physiol. 80, 429-434.

Hiller, R.G., Møller B.L. and Høyer-Hansen G. 1980. Characterization of six putative photosystem I mutants in barley. Carlsberg Res. Commun. 45, 315-328.

Høyer-Hansen, G., Bassi R., Hønberg L.S. and Simpson D.J. 1987. Immunological characterization of chlorophyll a/b-binding proteins of barley thylakoids. Planta, in press.

Ish-Shalom, D. and Ohad, I. 1983. Organization of chlorophyll-protein complexes of photosystem I in Chlamydomonas reinhardii. Biochim. Biophys. Acta 722, 498-507.

Machold, O. and Meister A. 1978. Resolution of the light harvesting chlorophyll-a/b protein of Vicia faba chloroplasts into two different chlorophyll-protein complexes. Biochim. Biophys. Acta 546, 472-480.

Melis, A. 1986. Functional properties of photosystem II_β in spinach chloroplasts. Biochim. Biophys. Acta 808, 334-342.

Michel, H. and Deisenhofer J. 1986. X-ray diffraction studies on a crystalline bacterial photosynthetic reaction center: a progress report and conclusions on the structure of photosystem II reaction centers. In: "Encyclopedia of Plant Physiol., new series) (Eds. L.A. Staehelin and Arntzen) (Springer, Berlin). Vol. 19, 371-381.

Miller, K.R. and Cushman R.A. 1979. A chloroplast membrane lacking photosystem II. Thylakoid stacking in the absence of the photosystem II particle. Biochim. Biophys. Acta 546, 481-497.

Mullet, J.E., Burke J.J. and Arntzen C.J. 1980. Chlorophyll-proteins of photosystem I. Plant Physiol. 65, 814-822.

Namba, O. and Satoh K. 1987. Isolation of a photosystem II reaction center consisting of D1 and D2 polypeptides and cytochrome b-559. Proc. Natl. Acad. Sci. USA 84, 109-112.

Olive, J., Wollman, F.A., Bennoun, P. and Recouvreur, M. 1979. Ultrastructure - function relationship in Chlamydomonas reinhardtii thylakoids, by means of a comparison between the wild type and the F_{34} mutant which lack the photosystem II reaction center. Molec. Biol. Rep. 5, 139-143.

Olive, J., Wollman, F.A., Bennoun, P. and Recouvreur, M. 1983. Localization of the core and peripheripheral antennae of photosystem I in the thylakoid membranes of Chlamydomonas reinhardtii. Biol. Cell. 48, 81-84.

Sane, P.V., Goodchild, D.J. and Park, R.B. 1970. Characterization of photosystem I and II separated by a non-detergent method. Biochim. Biophys. Acta 216, 162-178.

Simpson, D.J. 1979. Freeze-fracture studies on barley plastid membranes. III. Location of the light-harvesting chlorophyll-protein. Carlsberg Res. Commun. 44, 305-336.

Simpson, D.J. 1981. The ultrastructure of barley thylakoid membranes. In: Photosysthesis III. Structure and Molecular Organization of the Photosynthetic apparatus. (Ed. G. Akoyunoglou), (Balaban Int. Sci. Services, Philadelphia, Pa.). pp. 15-22.

Simpson, D.J. 1982. Freeze-fracture studies on barley plastid membranes. V. Viridis-n^{34}, a photosystem I mutant. Carlsberg Res. Commun. 47, 215-225.

Simpson, D.J. 1983. Freeze-fracture studies on barley plastid membranes. VI. Location of the P700 chlorophyll a-protein 1. Eur. J. Cell Biol. 31, 305-314.

Simpson, D.J., Høyer-Hansen G., Chua N.-H. and von Wettstein D. 1977. The use of gene mutants in barley to correlate thylakoid polypeptide composition with the structure of the photosynthetic membrane. In: Photosynthesis 77. (Eds. D.O. Hall, Coombs J., Goodwin T.W.) (The Biochemical Society, London). pp. 537-548.

Simpson, D.J. and Robinson S.P. 1984. Freeze-fracture ultrastructure of thylakoid membranes in chloroplasts from manganese-deficient plants. Plant Physiol. 74, 735-741.

Simpson, D.J. and Andersson B. 1986. Extrinsic polypeptides of the chloroplast oxygen evolving complex constitute the tetrameric ESs particles of higher plant thylakoids. Carlsberg Res. Commun. 51, 467-474.

Vallon, O., Wollman F.A. and Olive J. 1985. Distribution of intrinsic and extrinsic subunits of the PSII protein complex between appressed and non-appressed regions of the thylakoid membrane: An immunocytochemical study. FEBS Lett. 183, 245-250.

Vallon, O., Wollman F.A. and Olive J. 1986. Lateral distribution of the main protein complexes of the photosynthetic apparatus in Chlamydomonas reinhardtii and in spinach: an immunocytochemical study using intact thylakoid membranes and a PSII-enriched membrane preparation. Photobiochem. Photobiophys. 12, 203-220.

Vallon, O., Høyer-Hansen G. and Simpson D.J. 1987a. Photosystem II and cytochrome b-559 in the stroma lamellae of barley chloroplasts. Carlsberg Res. Commun. 52, in press.

Vallon, O., Wollman F.A. and Olive J. 1987b. Immunocytochemical studies on the organization of thylakoid membrane components. In: Progress in Photosynthesis Research, (Ed. J. Biggins), (Martinus Nijhoff Publ., Dordrecht, Netherlands). Vol II, 329-332.

Yamagishi, A. and Katoh S. 1985. Further characterization of two photosystem II reaction center complex preparations from the thermophilic cyanobacterium Synechococcus sp. Biochim. Biophys. Acta 807, 74-80.

PHOSPHORYLATION PROCESSES INTERACTING IN VIVO IN THE THYLAKOID MEMBRANES FROM C. REINHARDTII

F.-A. Wollman, C. Lemaire
Service de Photosynthèse
Institut de Biologie Physico-Chimique
13, rue Pierre et Marie Curie
75005 Paris, France

ABSTRACT

Phosporylation of thylakoid membrane proteins which occurs in vivo in C. reinhardtii, cannot be accounted for by a single kinase activity. We consider the existence of two distinct kinases which differ in substrate specificity: an activable LHC-kinase, the activity of which is controlled by the b6/f complex, and a continuously active PSII-kinase which is LHC-dependent. A model is proposed which involves an interaction between three groups of phosphoproteins. This interaction could arise either from a phosphotransferase process or from a sequence of substrate-induced modifications.

Higher plants and algae have developped an adaptation mechanism to cope with unbalanced excitation of the two photosystems. This mechanism, first described as producing State transitions (Bonaventura and Myers, 1969), corresponds to a change in antenna size of each photosystem: the antenna size of the less excited photosystem increases at the expense of that of the overexcited photosystem.

Since the two photosystems are located in two different membrane domains, the grana for PSII and the stroma lamellae for PSI, it is likely that the above regulation involves antenna movement along the thylakoid membrane. It has been shown that this "mobile antenna" belongs to the LHC (Larsson et al., 1983), a family of chlorophyll-binding proteins originally described as a PSII peripheral antenna (Thornber, 1975). Some of the LHC subunits undergo reversible phosphorylation upon state transitions (Bennett, 1979). It has been proposed that the driving force for LHC migration out of the grana arises from electrostatic repulsion between negatively charged groups, among which are the phosphate groups bound to some LHC subunits (Staehelin and Arntzen, 1983).

The mechanism by which reversible phosphorylation occurs is a central part in our understanding of the overall regulation process. It is currently described by the interplay of a continuously active phosphatase and a light-activated kinase (Allen et al., 1981). The activation of the latter enzyme can be produced in experimental conditions where the intersystem electron carriers are in a reduced state. Indirect evidence have favored kinase activation by the reduced plastoquinones (Allen et al., 1981)

Thylakoid membranes from higher plants and green algae contain several distinct phosphopolypeptides. For instance, in C. reinhardtii, we detected phosphorylation on five subunits of the peripheral antenna and four PSII subunits (Wollman and Delepelaire, 1984). The question then arises as to whether this phosphorylation process involves a single membrane-bound kinase. There is increasing evidence for multiple kinase activities in the thylakoid membranes from higher plants. At least two kinases have been purified from spinach thylakoid membranes (Lin et al., 1982, Coughlan and Hind, 1986). Kinetics of dephosphorylation of a 10kDa subunit of PSII are different from those of LHC (Steinback et al., 1982). Selective inhibition of phosphorylation of either PSII or LHC can be acheived in a variety of ways (Farchaus et al. 1985, Millner et al., 1982).

In the course of our study of the changes in polypeptide phosphorylation which occur in vivo in the green algae C. reinhardtii, we made several observations which support the existence of inter-related phosphorylation processes in the thylakoid membranes:
- the redox control of PSII phosphorylation is somewhat erratic in contrast with that of LHC, which show increased phosphorylation in reducing conditions. In particular the PSII subunit D2 often shows overphosphorylation in oxidizing conditions (Delepelaire and Wollman, 1985).
-based on kinetic studies of the phosphorylation changes, we distinguished two groups of phosphopolypeptides within the five subunits of the peripheral antenna which participate in the regulation process. LHCa, consisting of three subunits M.W. 35, 33 and 30.5kDa, displayed (de)phosphorylation kinetics which

matched those of the state transitions, at variance with the kinetics of LHCb, a group of two subunits, M.W. 29 and 25kDa (Delepelaire and Wollman, 1985).
-in the absence of the b6/f complex, phosphorylation of LHCb is totally abolished whereas that of LHCa still occurs (Lemaire et al., 1986). However the latter phosphorylation remains low and is no longer redox-controlled. On the other hand, phosphorylation of PSII subunits is unaltered. Mutants lacking the b6/f complex display such a phosphorylation pattern and do not undergo state transitions.

The tentative model presented below is based on these observations and takes into account two other sets of data:
- mutants lacking in PSII undergo larger phosphorylation changes upon state transitions than the WT strain (Wollman and Delepelaire, 1984).
- the absence of LHCa and LHCb prevents phosphorylation of the PSII subunits (Wollman and de Vitry, unpublished observation).

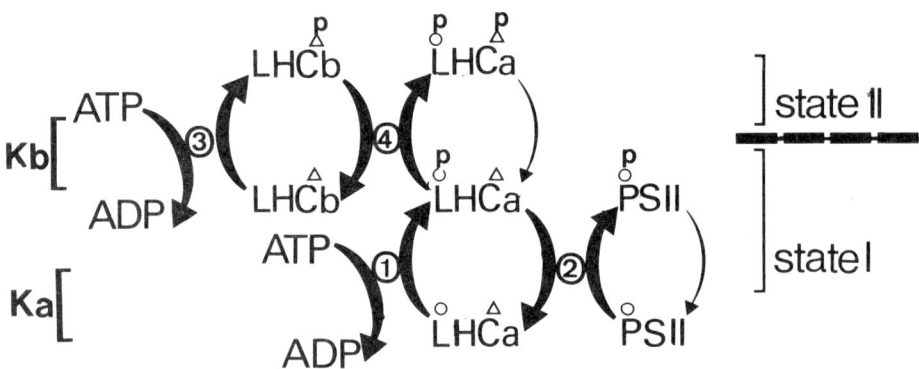

In this model, there are three groups of target polypeptides, LHCa, LHCb and PSII, for two membrane-bound kinases Ka and Kb. A characteristic of the model is that the phosphorylation status of either group of phosphopolypeptides influences that of the others. Here, this interaction is described as a phosphotransferase process, by analogy with the PEP/sugar PTS in bacteria (reviewed in Dills et al., 1980). An alternative view, which accounts equally well for the data, involves a sequence of substrate-induced modifications. In this case, reactions 3 and 4 for instance, would be expressed as follows:

$$\overset{P}{\underset{}{L}HCa} + L\overset{\triangle}{HCb} \xrightarrow[Kb]{③} \overset{P}{\underset{}{L}HCa} * L\overset{P}{\underset{}{H}Cb} \xrightarrow[Kb]{④} \overset{P}{\underset{}{L}HCa} + L\overset{P}{\underset{}{H}Cb}$$

The two kinases differ in redox control and substrate specificities. Ka shows continuous activity and is responsible for the phosphorylation pattern observed in thylakoid membranes from the b6/f mutants. It is specific for the phosphorylation of the PSII subunits. These subunits are substrate for Ka only when interacting with LHCa. Kb corresponds to the "LHC-kinase" involved in state transitions. It is specific for LHCb phosphorylation which is a prerequisit for further phosphorylation of LHCa by Kb. Kb activity depends on the presence of b6/f complexes in a way which may involve activation by cytbh. This cytochrome has a midpoint potential of 0mv (Joliot and Joliot, 1987) consistent with that required for Kb activation (Horton et al., 1981). It is worth noting that such a heme-regulated kinase is involved in the initiation of translation (Edelman et al., 1987). In the latter case, SH groups are involved in the regulation of the enzyme activity, a feature that has also been recognized for the LHC-kinase (Millner et al., 1982).

Further characterization of Ka and Kb is needed to establish whether these are the functional counterparts of the 24 and 64kDa proteins showing kinase activity, which have been purified from spinach thylakoid membranes (Lin et al., 1982; Coughlan and Hind, 1986).

REFERENCES

Allen, J.F., Bennett,J., Steinback,K.E. and Arntzen,C.J. 1981. Chloroplast protein phosphorylation couples plastoquinone redox changes to distribution of excitation energy between the two photosystems. Nature (Lond) 291, 25-29.
Bennett, J. 1979. Chloroplast phosphoproteins. Eur. J. Biochem., 99, 133-137.
Bonaventura, C. and Myers, J. 1969. Fluorescence and oxygen evolution from C. pyrenoidosa. Biochim. Biophys. Acta, 189, 366-383.
Coughlan, S.J. and Hind, G. 1986. Protein kinases from the thylakoid membrane. J. Biol. Chem., 261, 14062-14068.
Delepelaire, P. and Wollman, F.-A. 1985. Correlations between fluorescence and phophorylation changes in thylakoid membranes from C. reinhardtii. Biochim. Biophys. Acta, 809, 277-283.
Dills, S.S., Apperson, A., Schmidt, M.R. and Saier, M.H. 1980. Carbohydrate

transport in bacteria. Microbiol. Rev., 44, 385-418.

Edelman, A.M., Blumenthal, D.K. and Krebs, E.G. 1987. Serine/threonine kinases. Ann. Rev. Biochem., 56, 567-613.

Farchaus, J., Dilley, R.A. and Cramer, W.A. 1985. Selective inhibition of the spinach thylakoid LHCII protein kinase. Biochim. Biophys. Acta, 809, 17-26.

Horton, P., Allen, J.F., Black, M.T. and Bennett, J. 1981. Regulation of phosphorylation of chloroplast membrane polypeptides by the redox state of the plastoquinones. FEBS Lett., 125, 193-196.

Joliot, P. and Joliot, A. 1987. The low potential electron transfer chain in the cytochrome b/f complex. Biochim. Biophys. Acta, in press.

Larsson, U.K., Jergil, B. and Andersson, B. 1983. Changes in the lateral distribution of the LHC induced by phosphorylation. Eur. J. Biochem., 136, 25-29.

Lemaire, C., Girard, J. and Wollman, F.-A. 1986. Characterization of the b6/f complex subunits and studies on the LHC-kinase in C. reinhardtii using mutants altered in the b6/f complex. in Progress in Photosynthesis research, Vol. IV, pp. 655-658. Martinus Nijholoff eds. The Netherlands.

Lin, Z.F., Lucero, H.A. and Racker, E. 1982. Protein kinases from spinach chloroplast. J. Biol. Chem., 257, 12153-12156.

Millner, P.A., Widger, W.R., Abott, M.S., Cramer, W.A. and Dilley, R.A. 1982. The effect of adenine nucleotide on inhibition of the thylakoid kinase by SH-directed reagents. J. Biol. Chem., 257, 1736-1742.

Staehelin, L.A. and Arntzen, C.J. 1983. Regulation of chloroplast membrane function. J. Cell Biol., 97, 1327-1337.

Steinback, K.E., Bose, S. and Kyle, D.J. 1982. Phosphorylation of LHC regulates excitation energy distribution between PSII and PSI. Arch. Biochem. Biophys., 216, 356-361.

Thornber, J.P. 1975. Chlorophyll proteins. Ann. Rev. Plant Physiol., 26, 127-158.

Wollman, F.-A. and Delepelaire, P. 1984. Correlation between changes in light energy distribution and changes in thylakoid membrane polypeptide phosphorylation in C. reinhardtii. J. Cell Biol., 98, 1-7.

ISOLATION AND SOME PROPERTIES OF PHOTOSYNTHETIC MEMBRANE VESICLES ENRICHED IN PHOTOSYSTEM I FROM THE CYANOBACTERIUM *PHORMIDIUM LAMINOSUM* BY A NON-DETERGENT METHOD

J.L. Serra, J.A.G. Ochoa de Alda and M.J. Llama
Departamento de Bioquímica y Biología Molecular
Facultad de Ciencias
Universidad del País Vasco/Euskal Herriko Unibertsitatea
Apartado 644, E-48080 Bilbao, Spain

ABSTRACT

A relatively simple and quick procedure for the isolation of Photosystem I-enriched particles from the thermophilic cyanobacterium *Phormidium laminosum*, without the use of detergents for solubilization, is described. The procedure involves sonication of cells, centrifugation and DEAE-cellulose chromatography. The particles had an O_2 uptake activity of up to 200 μmol O_2 · mg chlorophyll^{-1} · h^{-1} and appeared as vesicles of 200 ±100 nm diameter when observed under electron microscopy. The analysis of the chlorophyll-protein complexes by polyacrylamide gel electrophoresis showed that these particles are enriched in the complexes associated with Photosystem I and partially depleted in those associated with Photosystem II. The particles did not contain ferredoxin and were active in NADP-photoreduction only in the presence of added ferredoxin. They were also able to photoreduce externally added electron mediators using ascorbate as electron donor; the reduced mediators can be coupled to hydrogenase for the production of H_2 or for the activation of cyanobacterial phosphoribulokinase using a ferredoxin/thioredoxin system.

INTRODUCTION

Thermophilic cyanobacteria, due to their increased heat stability and strong similarities in photosynthesis to that of higher plants, are one of the preferred group of microorganisms used for the separation and characterization of individual photosystems and of Chl-protein complexes. The most commonly used method for the isolation of photosynthetic membrane particles comprises detergent solubilization of membranes, molecular sieve or ion exchange chromatography, followed by sucrose density gradient centrifugation or electrophoresis. Detergents can radically alter the structure and functional properties of the Chl-protein complexes (Huang and Berns, 1983; Shiozawa et al., 1974; Markwell et al., 1980). However, it is generally thought that the purification of such complexes i nevitably requires the use of detergents (Thornber, 1979) and thus a great number of these tensoactive agents such as digitonin (Newman and Sherman, 1978;

Abbreviations: BSA, bovine serum albumin; Chl, chlorophyll; DCPIP, 2,6-dichlorophenol indophenol; DMBQ, 2,6-dimethyl-1,4-benzoquinone; DTT, dithiothreitol; Fd, ferredoxin; Hepes, N-2-hydroxyethyl-piperazine N'-2-ethane sulphonate; LDAO, lauryldimethyl amine oxide; Mes, 4-morpholine ethane sulphonic acid; MV, methyl viologen; PAGE, polyacrylamide gel electrophoresis; PRuK, phosphoribulokinase; PS, photosystem; Ru5P, ribulose 5-phosphate; SDS, sodium dodecyl sulphate; TMPD, N,N,N',N'-tetramethyl paraphenylene diamine; TRX, thioredoxin

Guikema and Sherman, 1983; Yamagishi and Katoh, 1983), sodium or lithium dodecyl sulphate (Öquist et al., 1981, Wollman and Bennoun, 1982) and the non ionic LDAO (Stewart, 1980; Bowes et al., 1983a) or Triton X-100 (Ogawa et al., 1969; Shiozawa, 1980), individually or in combination, (Nechustai et al., 1983; Bowes et al., 1983b; Stewart and Bendall, 1980), have been employed in solubilizing the photosynthetic membrane components of higher plants, algae and cyanobacteria.

A method has been reported for the isolation of six Chl a-protein complexes from the cyanobacterium *Phormidium luridum,* which did not involve the use of detergents (Huang and Berns, 1983). This non-detergent procedure included lysozyme-treatment of the cells followed by two chromatographic separations on Sepharose-4B and, finally two successive sucrose density gradient ultracentrifugations, the last one at 269,000 x g for 48 h. The main fraction recovered (F65, according to its position in the 20-65% sucrose gradient after centrifugation) was enriched in PSI while two more pale green fractions contained both PSI and PSII activities.

In this paper we describe an easy, reproducible, and fast procedure for obtaining PSI-enriched particles by a non-detergent method from the thermophilic cyanobacterium *Phormidium laminosum*. These particles showed a PSI activity (ascorbate/DCPIP → MV) of up to 200 µmole O_2 taken up per mg chlorophyll per h, and negligible PSII activity (H_2O → DMBQ or ferricyanide). We also present evidence on the feasibility of H_2 photoproduction and phosphoribulokinase photoactivation mediated by such PSI particles coupled to hydrogenase or to a Fd/TRX system.

MATERIALS AND METHODS

Chemicals. All chemicals used were of the highest available purity and were obtained from Sigma Chemical Co. London, and BDH Chemicals Ltd., Poole, Dorset, U.K. DEAE-cellulose (DE-52) was purchased from Whatman, Maidstone, U.K. and Sephacryl S-300 from Pharmacia Fine Chemicals, Uppsala, Sweden.

Strain and culture conditions. Phormidium laminosum (strain OH-1-p. Cl1) was originally obtained from Dr. R.W. Castenholz (University of Oregon, Eugene, OR, U.S.A.) and grown in medium D (Castenholz 1970) supplemented with $NaHCO_3$ (0.5 g · l^{-1}) at 45 °C. Cultures were vigorously stirred by an air stream (about 6 l · min^{-1}) which was bubbled through culture media. Cells were continuously illuminated with white light from 6 x 20 W fluorescent lamps (Tungsram F7 Daylight) and after 4-5 days of growth they were collected by centrifugation at 12,000 x g and washed with 50 mM Tris-HCl, pH 8.0.

Preparation of PSI-enriched particles. The cell suspension was diluted to about 0.25-0.30 mg Chl · ml^{-1} and sonicated, in an ice bath, in 40 ml batches in 10 s on/off intervals for 20 min with an MSE sonifier at medium power, 10 micron setting or in 5 ml batches with a Dawe Soniprobe 4 amp output, 15 s on/off for 5 min. The broken material was centrifuged at 40,000 x g for 20 min and the pellet discarded. Sufficient dry DEAE-cellulose (about 10 g DE-52 per 100 ml of preparation) was added to the blue supernatant to ensure that all the blue material (phycobiliproteins) bound to the ion exchanger and after mild shaking for a few minutes the slurry was poured into a chromatography column. The unbound material, which appeared as a green eluate, was collected by hydrostatic pressure until the column was completely drained. The eluate was concentrated either by treating with polyethyleneglycol 6000 or by ultrafiltration in a Diaflo cell (Amicon) fitted with a PM-10 membrane at 20 psi N_2 pressure. Further purification was achieved by gel filtration of the concentrated material through a Sephacryl S-300 column equilibrated and developed with 50 mM Tris-HCl, pH 8.0, containing 50 mM KCl. After the sonication step the preparations were kept covered with aluminium foil to minimise exposure to light and all operations were carried out in a cold room at 4 °C.

Activity measurements. Photosynthetic O_2 exchange was measured using an O_2 electrode (Rank Brothers, Bottisham, U.K.) illuminated with saturating white light (360 µE · m^{-2} · s^{-1}) according to Stewart and Bendall (1980) except that azide was substituted for catalase in PSI assay and 6 mM DMBQ was used as electron acceptor of PSII.

Hydrogenase activities were measured using MV chemically reduced with sodium dithionite or photochemically with PSI particles. Photoproduction pf H_2 was carried out in 15 ml glass vials under a nitrogen atmosphere, at 30 °C, in a shaking water bath illuminated with white light of intensity 250 µE · m^{-2} · s^{-1}. H_2 evolved was determined gas chromatographically (Rao et al., 1978).

Phosphoribulokinase activity was measured by the spectrophotometric method with pyruvate kinase and lactate dehydrogenase as auxillary enzymes (Hurwitz et al., 1956). Photoactivation of the purified *Anabaena cylindrica* PRuK was assayed at 25 °C in 15 ml glass vials fitted with subba seal stoppers. The vessels contained in a final volume of 1.5 ml, 100 mM Hepes-NaOH buffer (pH 7.5), 25 µM *Spirulina maxima* Fd, 0.4 µM *A. cylindrica* TRX, enriched-PSI particles from *P. laminosum* equivalent to 36 µg of Chl, 100 µM DCPIP, 10 mM sodium ascorbate, 1 mM $MgCl_2$, 2 mM TMPD and 0.5 ml of the Fd-TRX reductase preparation. After flushing for 10 min with N_2, vessels were preincubated for 10 min in the light (about 250 µE · m^{-2} · s^{-1}). The activation reaction was started by injecting PRuK into the vessels and was continued for 30 min under the same conditions. Control vessels wrapped with aluminium foil and incubated in the same conditions, or in the absence of Fd-TRX reductase were included. At zero time and at 10 min intervals aliquots of 0.4 ml were withdrawn by a syringe and injected into a cuvette containing 0.6 ml of the complete standard reaction mixture used for the spectrophotometric assay of PRuK (containing 2.0 µmol of Ru5P) and the decrease of $A_{340\,nm}$ was recorded at 30 °C for 10 min.

Electrophoresis conditions. Samples were prepared for electrophoresis as follows. Membrane vesicles [2 mg Chl · ml^{-1} in 25 mM Tris-HCl, pH 8.0, and 25% (v/v) glycerol] were first pretreated with 3.5% (v/v) LDAO in buffer C (Stewart and Bendall, 1979) at a final ratio LDAO: Chl = 3.5: 1. After 30 min at 4 °C in the dark, they were treated with 10% (w/v) SDS in buffer C at a final ratio SDS: Chl = 20: 1, and were used immediately for electrophoresis. Electrophoresis of Chl-protein complexes was carried out on 1 mm-thick slab gels [3% (w/v) acrylamide stacking gel and 5% (w/v) acrylamide resolving gel] in a PhastSystem (Pharmacia, Uppsala) equipment. The buffer system in the stacking and in the resolving gels were 60 mM Tris-HCl (pH 6.7) and 375 mM Tris-HCl (pH 8.9), respectively. The buffer system in SDS buffer 2% (w/v) agarose strips was 0.2 M Tricine, 0.2 M Tris and 0.55 (w/v) SDS, pH 7.5. Electrophoresis was carried out at 4 °C for 30-60 min at a constant voltage of 100 V. Chlorophyll distribution was determined by scanning the unstained gels at 672 nm on a Shimadzu CS-930 scanner.

Analytical methods. Chlorophyll was estimated after extraction into 80% (v/v) acetone according to Arnon et al., (1974). Protein was determined according to Wang and Smith (1975) using crystalline bovine serum albumin as standard.

RESULTS AND DISCUSSION

A typical preparation of PSI-enriched particles by the non-detergent method is summarized in Table 1. After DEAE-cellulose chromatography the preparation showed a specific activity of more than 130 µmol O_2 · mg $Chl^{-1} \cdot h^{-1}$, with a recovery of PSI activity near to 30%. In some preparations PSI activity up to near 200 µmol O_2 · mg $Chl^{-1} \cdot h^{-1}$ was measured. This activity could be increased two- to three-fold in the presence of 1% (v/v) Triton X-100 (data not shown).

TABLE 1. Purification of PSI-enriched particles from *Phormidium laminosum*

Purification step	Protein (mg)	Chl a (mg)	PSI µmol O_2/h	Specific activity µmol O_2/h		Purification (-fold)		Yield on PSI activity (%)
				per mg Chl	per mg prot	Chl basis	prot basis	
Sonicate	832	26	690	26.5	0.83	1.0	1.0	100
Super 40 000 xg	370	7	353	50.4	0.95	1.90	1.14	51
Pellet 40 000 xg	430	18	267	14.8	0.62	0.56	0.75	38.7
DE-52 eluate	28.3	1.43	190	132.8	6.71	5.0	8.1	27.5

The absorption spectra of the fractions obtained during the purification of the PSI particles are shown in Fig 1a. Both the sonicate and the original 40,000 x g supernatant showed a major peak at 620 nm due to soluble phycobiliproteins, which can be completely removed by batch chromatography on DEAE-cellulose. Thus, the A_{620}/A_{676} ratio, which compares the absorption maxima of blue pigments to that of chlorophyll was 2.73 in the sonicate, 3.81 in the 40,000 x g supernatant, and only 0.27 in the DE-52 eluate.

When the concentrated DE-52 eluate was chromatographed on Sephacryl S-300 the specific activity of particles remained unchanged on a chlorophyll basis but increased on the basis of protein due to the removal of some minor contaminant proteins. This molecular sieve also suggested that the photosynthetic particles were nearly homogeneous with respect to size since the green material was eluted as a sharp peak immediately before the thyroglobulin marker (M_r 669 kDa) but after the void volume calculated with a Blue Dextran standard. The absorption spectrum of this fraction (Fig. 1b) showed major peaks at 676 and 437 nm and shoulders at 422 and 380 nm corresponding to chlorophyll a, and a broad shoulder at 485 nm due to the

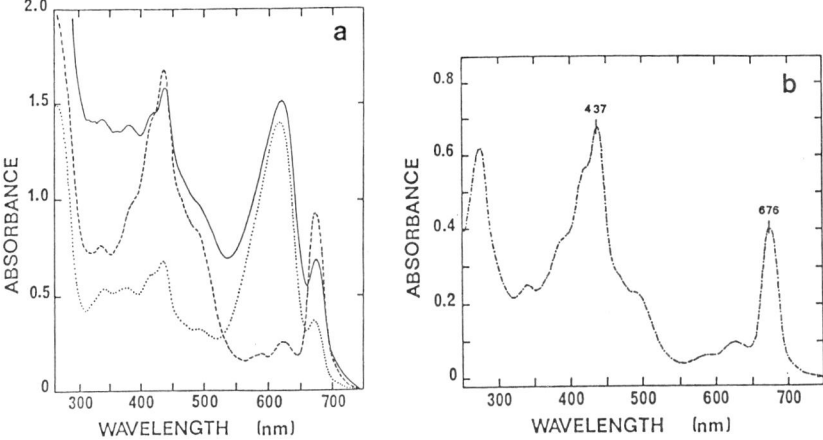

Fig. 1. Absorption spectra of the different fractions obtained during the purification of PSI-enriched particles of *P. laminosum*. The spectra were recorded at room temperature on a Beckman UV 5260 spectrophotometer using 1 cm path length quartz cuvettes. (a) Sonicate (——); 40,000 x g supernatant (· · ·); DEAE-cellulose eluate (- - -). (b) PSI-enriched particles after chromatography on a Sephacryl S-300 column.

presence of carotenoids. In the ultraviolet region two peaks at 340 and 276 nm can be observed. The A_{437}/A_{676}, A_{437}/A_{280}, A_{676}/A_{280} and A_{620}/A_{676} ratios were 1.68, 1.13, 0.67 and 0.23, respectively. The absorption spectrum of purified PSI-enriched particles shown in Fig. 1b is quite similar to that reported for PSI-enriched particles from *P. luridum* isolated with SDS (Thornber, 1969) or from *P. laminosum* isolated with LDAO and electrophoretically separated in the presence of SDS (Stewart, 1980).

Chemical difference spectra (ascorbate-reduced *minus* ferricyanide-oxidized) showed the presence of P700 in these particles. The non-detergent-treated particles were unable to penetrate into 6% polyacrylamide gels, and all the chlorophyllous material was retained at the top of the gel after electrophoresis, indicating the absence of free Chl-proteins in these particles.

Negative-staining electron micrographs of our PSI-enriched particles (Fig. 2) show more-or-less spherical vesicles of 200 ±100 nm in diameter.

In order to elucidate whether the apparent enrichment in PSI activity and the concomitant decrease in PSII activity showed by the photosynthetic particles obtained from the DE-52 eluate was due to a loss of specific Chl-protein complexes rather than an inactivation of PSII, the composition of such Chl-protein complexes was examined by SDS-PAGE. Fig. 3 shows the electrophoretic pattern obtained for complete membrane fragments, isolated and treated as reported by Stewart and Bendall (1980), and for PSI-enriched particles of *P. laminosum*. The four green bands, corresponding to bands A, A-1, A-2 and FP reported for membrane

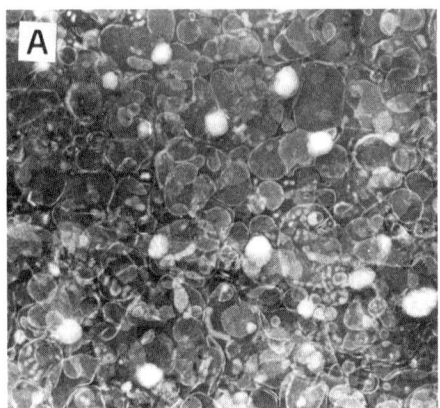
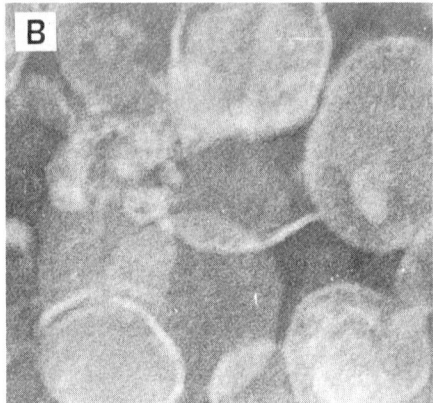

Fig. 2. Micrographs of PSI-enriched particles from *P. laminosum* after negative staining with 2% (w/v) phosphotungstic acid. Magnification were: (A) 25,000x and (B) 160,000x.

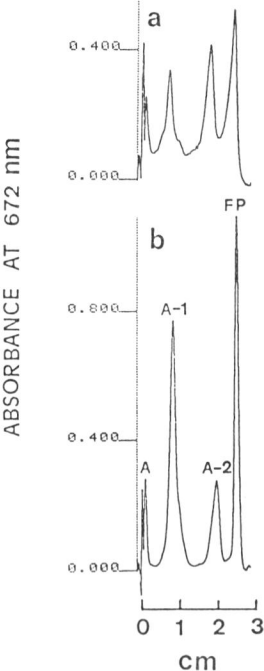

Fig. 3. Densitometer scans showing electrophoretic separation of Chl-protein complexes in *P. laminosum*. Samples were pretreated with LDAO at a ratio LDAO: Chl = 3.5: 1 and treated with SDS at a ratio SDS: Chl = 20: 1 as described under Materials and Methods. (a) membrane fragments isolated according to Stewart and Bendall (1980), (b) PSI-enriched particles. The area of the peaks corresponding to the free pigment (FP) bands was similar in both cases.

fragments of *P. luridum* (Reinman and Thornber, 1979), were also observed in both cases, although they were present in different relative ratios.

Bands A and A-1 are associated with PSI, with the former being probably an oligomeric form of the latter whereas A-2 are associated with PSII (Stewart, 1980). The area of peaks corresponding to the distinct green bands shown in Fig. 3 are as follows. Membrane fragments: (A + A-1) = 32%, A-2 = 35% and FP = 33%; PSI-enriched particles: (A + A-1) = 50%, A-2 = 21% and FP = 27%. These figures suggested that functionally PSI-enriched particles are also structurally enriched in the Chl-protein complexes associated with PSI and depleted in the Chl-protein complexes associated with PSII, when compared to the membrane fragments. However, the PSI-enriched particles still contained 21% of the total chlorophyll associated with PSII although the PSII activity was negligible in such preparations.

The detergent treatment conditions of samples drastically affected the SDS-PAGE of the Chl-protein complexes. Thus, when samples were SDS-treated for different time lengths the A-2 band was affected faster than the other green bands, and no A-2 band could be observed after incubation of samples for 90 min with SDS at a final SDS: Chl ratio = 20: 1 (data not shown). However, the effect of the LDAO pretreatment of samples was less apparent.

The photoproduction of H_2 using ascorbate as a substrate by PSI of chloroplasts (Muallem and Hall, 1982) or cyanobacterial membranes (Serra et al., 1986) has been previously reported. Rates of H_2 photoevolution catalysed by our PSI-enriched particles or hydrogenase entrapped in calcium alginate beads, either alone or in combination, are shown in Fig. 4. An improvement in the longevity of H_2 production as a result of

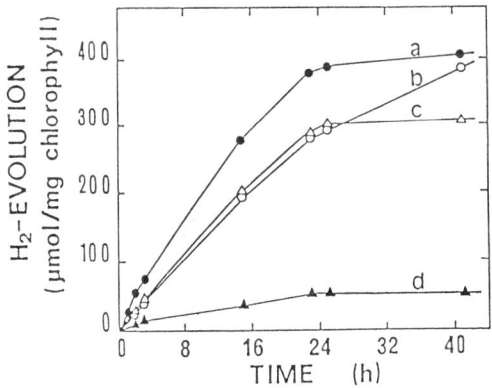

Fig. 4. H_2 photoevolution from ascorbate via *P. laminosum* PSI particles under different conditions of immobilization. The complete assay system contained 100 mM Mes-NaOH buffer, pH 7.0; 75 mM sodium ascorbate; 15 mM DTT; 2 mM TMPD; 1% (w/v) BSA; PSI particles (30 µg Chl); 50 µl (saturating amount) of *Clostridium pasteurianum* hydrogenase and 12.5 µM *Spirulina maxima* Fd as electron mediator. (a) Conditions: all components free; (b) hydrogenase immobilized in Ca alginate according to Gisby and Hall (1980); (c) PSI particles immobilized in Ca alginate; (d) hydrogenase and PSI coimmobilized in Ca alginate.

immobilization was observed only where the hydrogenase enzyme was entrapped in the alginate. Although after immobilization the PSI-enriched particles from *P. laminosum* retained their photosynthetic activity in H_2 evolution systems for much longer periods when compared to isolated chloroplast membranes, the particles were not as stable as immobilized whole cells of cyanobacteria.

Finally, in order to ascertain if our PSI-particles were also useful for enzyme activation by potochemically reduced TRX, the pure *A. cylindrica* PRuK was incubated with PSI-enriched particles, TRX and Fd-TRX reductase purified from *A. cylindrica* (Serra et al., 1987) and a number of natural and artificial electron mediators as described under Materials and Methods. As shown in Fig. 5, PRuK was activated by photoreduced TRX and the degree of activation achieved depended on the time of preincubation (compare curves b-d). No activation was observed when the enzyme was preincubated in the dark with the complete Fd/TRX system, or when the Fd-TRX reductase was omitted.

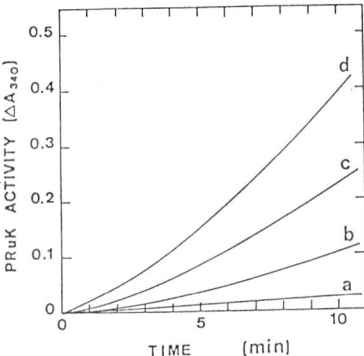

Fig. 5. Time course of the photoactivated *A. cylindrica* PRuK activity. The pure enzyme was photoactivated by preincubation in the light in the presence of TRX, Fd-TRX reductase, *P. laminosum* PSI particles and several electron donors and mediators, as described under Materials and Methods. Preincubation conditions: complete system either in the dark for any time or in the light at zero time, or in the absence of Fd-TRX reductase after 30 min in the light (a); complete system in the light for: 10 min, (b); 20 min, (c); 30 min, (d). The activity in (d) was 6.9 μmol of ribulose 1,5-bisphosphate formed · min^{-1} · mg protein^{-1}.

ACKNOWLEDGMENTS

This work has been partially supported by grants from the Universidad del País Vasco/Euskal Herriko Unibertsitatea (310.03-5/86), FEBS and the British Council.

REFERENCES

Arnon, D.I., McSwain, B.D., Tsujimoto, H.Y. and Wada, K. 1974. Photochemical activity and components of membrane preparations from blue-green algae. I. Coexistence of two photosystems in relation to chlorophyll a and removal of phycocyanin. Biochim. Biophys. Acta, 357, 231-245.
Bowes, J.M., Horton, P. and Bendall, D.S. 1983a. Characterization of photosystem II electron acceptors in *Phormidium laminosum*. Arch. Biochem. Biophys., 225, 353-359.
Bowes, J.M., Stewart, A.C. and Bendall, D.S. 1983b. Purification of photosystem II from *Phormidium laminosum* using the detergent dodecyl-β-D-maltoside. Biochim. Biophys. Acta, 725, 210-219.
Castenholz, R.W. 1970. Laboratory culture of thermophilic cyanophytes. Schweiz. Z. Hydrol., 32, 538-551.
Gisby, P.E. and Hall, D.O. 1980. Biophotolytic H_2 production using alginate-immobilized chloroplasts, enzymes and synthetic catalysts. Nature, 287, 251-252.
Guikema, J.A. and Sherman, L.A. 1983. Chlorophyll-protein organization of membranes from the cyanobacterium *Anacystis nidulans*. Arch. Biochem. Biophys., 220, 155-166.
Huang, C. and Berns, D.S. 1983. Partial characterization of six chlorophyll a-protein complexes isolated from a blue-green alga by a non-detergent method. Arch. Biochem. Biophys., 220, 145-154.
Hurwitz, J., Weissbach, A., Horecker, B.L. and Smyrniotis, P.Z. 1956. Spinach phosphoribulokinase. J. Biol. Chem., 218, 308-321.
Markwell, J.P., Thornber, J.P. and Skrdla, M.P. 1980. Effect of detergents on the reliability of a chemical assay for P-700. Biochim. Biophys. Acta, 591, 391-399.
Muallem, A. and Hall, D.O. 1982. Ascorbate as a substrate for photoproduction of hydrogen by photosystem I of chloroplasts. Plant Physiol., 69, 1116-1120.
Nechushtai, R., Muster, P., Binder, A., Liveanu, V. and Nelson, N. 1983. Photosystem I reaction center from the thermophilic cyanobacterium *Mastigocladus laminosus*. Proc. Acad. Sci. U.S.A., 80, 1179-1183.
Newman, P.J. and Sherman, L.A. 1978. Isolation and characterization of the photosystem I and II membrane particles from the blue-green alga *Synechococcus cedrorum*. Biochim. Biophys. Acta, 503, 343-361.
Ogawa, T., Vernon, L.P. and Mollenhauer, H.H. 1969. Properties and structure of fractions prepared from *Anabaena variabilis* by the action of Triton X-100. Biochim. Biophys. Acta, 172, 216-229.
Öquist, G., Fork, D.C., Schoch, S. and Malmberg, G. 1981. Solubilization and spectral characteristics of chlorophyll-protein complexes isolated from the thermophilic blue-green alga *Synechococcus lividus*. Biochim Biophys. Acta, 638, 192-200.
Rao, K.K., Gogotov, I.N. and Hall, D.O. 1978. Hydrogen evolution by chloroplast-hydrogenase systems: improvements and additional observations. Biochimie, 60, 291-296.
Reinman, S. and Thornber, J.P. 1979. The electrophoretic isolation and partial characterization of three chlorophyll-protein complexes from blue-green algae. Biochim. Biophys. Acta, 547, 188-197.
Serra, J.L., Llama, M.J., Rao, K.K. and Hall, D.O. 1986. B-5 Hydrogen photoproduction using photosystem I-enriched particles from *Phormidium laminosum*. Book of Abstracts 6th International Conference on Photochemical Conversion and Storage of Solar Energy, Lab. Biophysique, INSERM, Paris.
Serra, J.L., Llama, M.J., Rowell, P. and Stewart, W.D.P. 1987. Thioredoxin activation of *Anabaena* phosphoribulokinase, unpublished results.
Shiozawa, J.A., Alberte, R.S. and Thornber, J.P. 1974. The P700-chlorophyll a-protein. Isolation and some characteristics of the complex in higher plants. Arch. Biochem. Biophys., 165, 388-397.
Shiozawa, J.A. 1980. The P700-chlorophyll a-protein of higher plants. Methods Enzymol., 69, 142-150.
Stewart, A.C. 1980. The chlorophyll-protein complexes of a thermophilic blue-green alga. FEBS Lett., 114, 67-72.
Stewart, A.C. and Bendall, D.S. 1980. Photosynthetic electron transport in a cell-free preparation from the thermophilic blue-green alga *Phormidium laminosum*. Biochem. J., 188, 355-361.
Thornber, J.P. 1969. Comparison of a chlorophyll a-protein complex isolated from a blue-green alga with chlorophyll-protein complexes obtained from green bacteria and higher plants. Biochim. Biophys. Acta, 172, 230-241.
Wang, C.S. and Smith, R.L. 1975. Lowry determination of protein in the presence of Triton X-100. Anal. Biochem., 63, 414-417.

Wollman, F.A. and Bennoun, P. 1982. A new chlorophyll-protein complex related to photosystem I in *Chlamydomonas reinhardii*. Biochim. Biophys. Acta, 680, 352-360.

Yamagishi, A. and Katoh, S. 1983. Two chlorophyll-binding subunits of the photosystem 2 reaction center complex isolated from the thermophilic cyanobacterium *Synechococcus* sp. Arch. Biochem. Biophys., 225, 836-846.

OUTDOOR CULTURE OF SELECTED NITROGEN-FIXING BLUE-GREEN ALGAE FOR THE PRODUCTION OF HIGH-QUALITY BIOMASS

M.G. Guerrero, A.G. Fontes, J. Moreno, J. Rivas,
H. Rodríguez-Martínez, M.A. Vargas and M. Losada
Instituto de Bioquímica Vegetal y Fotosíntesis
Universidad de Sevilla y C.S.I.C.
Facultad de Biología, Apdo.1113
41080 Sevilla, Spain

ABSTRACT

Filamentous blue-green algae (cyanobacteria) with heterocysts are unique organisms in that they carry out oxygenic photosynthesis and N_2 fixation simultaneously. The lack of requirement of nitrogen fertilizer for vigorous growth represents a sensible economical advantage as well as a restriction to contamination by other organisms. The filamentous nature of these algae facilitates harvesting, another major problem in the technology of algal biomass production.

A screening of different strains of N_2-fixing blue-green algae has been undertaken according to criteria of biomass quality and productivity, as well as of tolerance to temperature, pH, illumination and salinity extremes. Optimization of conditions for outdoor culture of the selected strains has been attempted. Productivity values in the order of 20 g (dry weight) m^{-2} day^{-1} have been obtained for three different strains of Anabaena. Special attention deserves an Anabaena strain which contains high amounts of phycoerythrin, a microalgae pigment with interesting applications as coloring agent and as fluorescence tracer in diagnostics and biomedical research.

INTRODUCTION

Algal biomass has since long been considered an alternative source of protein that could supplement conventional food and feed production and help to lower the current protein deficit in many countries. Other practical uses of microalgal cultures include the reclamation of wastewater and the production of a variety of commercial chemicals on a renewable basis (Richmond, 1986). The biomass of some strains of microalgae is particularly rich in special lipids which can find applications in lubricants, drying oils, and substitutes for specialty natural waxes and oils in short supply. Special polysaccharides with a commercial value are also present as a part of the biomass of a number of species. A variety of pigments found in microalgae, including carotenes, xantophylls and phycobiliproteins, have a value as natural dyes in the food/feed industry (Benemann and Weissman, 1984; Benemann et al., 1987). Phycobiliproteins -photosynthetic antenna pigments unique to microalgae- have recently found additional applications in fluorescence immunoassays and fluorescence microscopy for diagnostics and biomedical research. (Glazer and Stryer, 1984; Biomeda, 1986).

Among the different groups of microalgae, the nitrogen-fixing blue-greens appear particularly attractive for the production of high-quality biomass, since they are able to synthesize all their cell components from water, air and a few mineral salts at the expense of solar energy. Nitrogen fertilizer, expensive in terms of money and energy, is not required as a component of the growth medium for these organisms, since they can use atmospheric nitrogen as the sole nitrogen source. The group of heterocystous filamentous blue-green algae is particularly relevant, since their representatives can perform effective N_2 fixation under aerobic conditions. The filamentous nature of these algae represents an advantage for harvesting of the algal cells. The lack of fixed nitrogen in the growth medium, aside from its economical implication, restricts the problem of contamination by other organisms. Despite these clear advantages, little has been done with regard to the establishment of cultures of N_2-fixing blue-green algae on a large scale (Fontes et al., 1983; 1987).

This work deals with the selection of strains of N_2-fixing blue-green algae on the basis of biomass composition, productivity and tolerance to temperature and pH. Outdoor culture of the most adequate strains has been tested in order to derive optimal conditions for growth.

RESULTS AND DISCUSSION

The most adequate strains with regard to biomass yield, cellular composition and tolerance to temperature, pH and high irradiance have been selected in laboratory experiments among the following nitrogen-fixing blue-green algae: Anabaena sp. strain ATCC 33047, Anabaena variabilis strain ATCC 29413,

Fischerella sp. strain PCC 7520, Nostoc sp. strain PCC 6719 and Nostoc sp. strain PCC 73102, from the American Type Culture Collection (ATCC) and the Pasteur Institute Culture Collection (PCC), as well as natural isolates of Anabaena sp., Anabaenopsis sp. and Nostoc paludosum from Albufera de Valencia, Spain (Fontes et al., 1983; Rodriguez-Martinez et al., 1987). The strains eventually chosen for further studies were A. variabilis ATCC 29413, Anabaena sp. ATCC 33047 and the Albufera isolate of Anabaena sp. The two latter are halotolerant species.

As an example of the data derived from the selection studies, Table 1 shows some interesting properties of the Albufera isolate of Anabaena sp. This strain exhibits broad optimal pH and temperature, together with a high productivity. As usual among the N_2-fixing blue-green algae so far tested, Anabaena sp. has a high protein content (about 60% crude protein). About one third of total protein corresponds to phycobiliproteins, accessory photosynthetic pigments specific of some microalgae groups. Particularly noteworthy, and uncommon among blue-greens, is the high C-phycoerythrin content found in Anabaena sp., amounting to more than 8% of total dry weight.

Phycoerythrin is a water-soluble pigment which exhibits red fluorescence with a high quantum yield, the emission being basically unaltered over broad temperature and pH range, and not modified by the presence of other biological compounds. Phycoerythrin moreover is stable and can be readily conjugated to other biomolecules without sensible alterations in light absorption and emission characteristics. These and other properties make phycoerythrin a candidate of choice for its use as fluorescent tracer (phycofluor) in a variety of immunoassays (Glazer and Stryer, 1984; Biomeda, 1986). The phycoerythrin--rich strain of Anabaena represents an excellent candidate for its use in the generation of algal biomass containing a high value product. Although the other strains of Anabaena tested have only low amounts of phycoerythrin (around 1%) they contain, however, substantial levels of C-phycocyanin (more than 10%) and allophycocyanin, likewise phycobiliprotein pigments of great interest.

TABLE 1   Some interesting properties of growth and composition of Anabaena sp. (Albufera isolate)

Productivity (g (d.w.) m^{-2} day^{-1})	40*
Optimum temperature (°C)	30-40
Optimum pH	6.2-8.8
Net protein (% of dry weight)	45
Phycobiliproteins (% of dry weight)	18.5
C-phycocyanin	7.1
Allophycocyanin	3.1
C-phycoerythrin	8.3

*This value corresponds to semicontinuous cultures in the laboratory at high irradiance (300 W m^{-2}, white light), under 12 h light/12 h dark cycles.

Anabaena variabilis, Anabaena sp. ATCC 33047 and Anabaena sp. (Albufera isolate) have been grown outdoors using two different culture systems, in which turbulence was alternatively provided by air-lifting from the bottom of the container or by a paddle-wheel. The paddle-wheel system has shown to be more effective, allowing better productivity and offering a lower cost than the air-lift option (Fontes et al., 1987; Vargas et al., 1985).

Using 1 m^2 containers provided with a paddle-wheel (Fig. 1) the effect of several factors on biomass yield have been assayed at 30°C in outdoor semicontinuous cultures. At the moderate irradiance proper of autumn and winter (below 10 MJ m^{-2} day^{-1}), the turbulence furnished by the paddle-wheel (15-20 rpm) suffices to provide all the C and N required for optimal cell growth. Under these conditions and in the presence of 50 mM $NaHCO_3$ but without any CO_2 supply (pH 9.3), productivity of Anabaena ATCC 33047 was 8-13 g (dry weight) m^{-2} day^{-1}. In summer (irradiance, 20-25 MJ m^{-2} day^{-1}), productivities of more than 25 g (dry weight) m^{-2} day^{-1} were recorded, provided that the cultures were supplemented with CO_2 at a flow rate of 1 mole m^{-2} day^{-1} in addition to bicarbonate. Under these conditions, addition of CO_2 was an absolute requirement for pH

Fig. 1 Container (1 m² surface) provided with a shaking paddle-wheel system for outdoor culture of nitrogen-fixing blue-green algae.

maintenance and growth enhacement, whereas a supply of air-N_2 or combined nitrogen did not improve productivity (Moreno et al., 1987).

In algae mass culture, selection of appropriate depth and cell density is of outmost importance for achieving high productivity. The results shown in Fig. 2 illustrate the effect of these factors on the biomass yield of Anabaena ATCC 33047. Increasing cell density over 2 mg (chl) l^{-1} or depth over 10 cm resulted in decreased productivity. These negative effects might well be in both cases consequence of mutual shading of the cells.

The conditions selected as optimal for outdoor growth of the selected strains of N_2-fixing blue-green algae were 10 cm depth and 20 rpm for the rotating speed of the paddle wheel. The optimal cell density varied along the year, according to

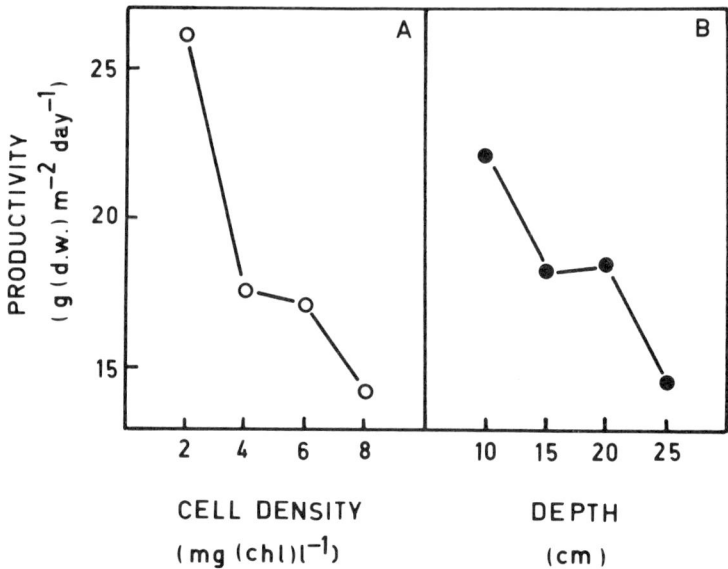

Fig. 2 Effect of cell density (A) and culture depth (B) on biomass productivity of Anabaena ATCC 33047 in semi-continuous culture outdoors. Culture depth in (A) was 10 cm and cell density in (B) was 2.5 mg (chl) l^{-1}. The culture medium (Moreno et al., 1987) was supplemented with 50 mM $NaHCO_3$ and CO_2 at a rate of 1-1.5 mol m^{-2} day^{-1}. The pH of the cell suspension was 9.1-9.4, and the temperature, 30°C. Once a day, part of the cell suspension was removed and replaced with fresh medium in order to maintain the selected cell density. The values shown are average of four independent determinations throughout four consecutive days in Summer, at a mean irradiance of 22 MJ m^{-2} day^{-1}.

seasonal irradiance, never exceeding the value of 5 mg (chl) l^{-1}. Under optimized conditions, productivities over 20 g (dry weight) m^{-2} day^{-1} have been obtained in summer for the three selected strains, namely A. variabilis, Anabaena sp. 33047 and Anabaena sp. (Albufera isolate).

The filamentous nature of the different Anabaena strains allows separation of cells from the medium by procedures other than the expensive centrifugation, such as settling and/or filtration. The Albufera isolate of Anabaena sp. used in this work exhibits a considerable advantage in this respect, since it easily autofloculates, especially at high pH values.

The results reported support the contention that N_2-fixing blue-green algae represent valuable candidates for their use in the generation of high quality biomass. High yields of cells with a biomass rich in proteins and some high value products can be achieved at the expense of sunlight energy by photosynthesis with water as the terminal reductant, air being the source of all the nitrogen and most of the carbon required for vigorous growth.

ACKNOWLEDGEMENTS

Research supported by grant BT 0045/85 from Comisión Asesora de Investigación Científica y Técnica. The authors thank Mrs. A. Friend and M.J. Pérez de León for skilful secretarial assistance.

REFERENCES

Benemann, J.R. and Weissman, J.C. 1984. Chemicals from microalgae. In "Bioconversion Systems" (Ed. D.L. Wise). (CRC Press, Boca Raton, Florida). pp. 59-70.

Benemann, J.R., Tillet, D.M. and Weissman, J.C. 1987. Microalgae biotechnology. Trends Biotechnol., $\underline{5}$, 47-53.

Biomeda, 1986. Phycobiliproteins as fluorescence tracers. Biomeda News $\underline{1}$, 1-12.

Fontes, A.G., Rivas, J., Guerrero, M.G. and Losada, M. 1983. Production of high-quality biomass by nitrogen-fixing blue-green algae. In "Energy from Biomass-2nd EC Conference" (Ed. A. Strub, P. Chartier and G. Schlesser). (Applied Science Publishers, London). pp. 265-269.

Fontes, A.G., Vargas, M.A., Moreno, J., Guerrero, M.G. and Losada, M. 1987. Factors affecting the production of biomass by a nitrogen-fixing blue-green alga in outdoor culture. Biomass $\underline{13}$, 33-43.

Glazer, A.N. and Stryer, L. 1984. Phycofluor probes. Trends Biochem. Sci., $\underline{9}$, 423-427.

Moreno, J., Vargas, M.A., Fontes, A.G., Guerrero, M.G. and Losada, M. 1987. Effect of several parameters on the productivity of a nitrogen-fixing blue-green alga in outdoor semicontinuous culture. In "Proceedings of the International Congress on Renewable Energy Sources" (Ed. S. Terol). (CSIC, Madrid), in press.

Richmond, A. (Ed.) 1986. Handbook of microalgal mass culture. (CRC Press, Boca Raton, Florida).

Rodríguez-Martinez, H., Rivas, J., Guerrero, M.G., Paneque, A. and Losada, M. 1987. Selection of nitrogen-fixing blue-green algae for biomass photoproduction. In "Proceedings of the International Congress on Renewable Energy Sources" (Ed. S. Terol). (CSIC, Madrid), in press.

Vargas, M.A., Moreno, J., Fontes, A.G., Guerrero, M.G. and Losada, M. 1985. Evaluation of two shaking systems for algal biomass photoproduction in outdoor semicontinuos cultures of Anabaena variabilis. In "Proceedings XXIII Congress Comples" (Eds. V. Ruíz and M.G. Guerrero). (ADESA, Sevilla). pp. 207-212.

HYDROGEN PEROXIDE PHOTOPRODUCTION BY BIOLOGICAL AND CHEMICAL SYSTEMS

M.A. De la Rosa, M. Roncel and J.A. Navarro
Instituto de Bioquímica Vegetal y Fotosíntesis,
Universidad de Sevilla y C.S.I.C.
Facultad de Biología, Apdo. 1095
41080 Sevilla, Spain

ABSTRACT

Several biological and chemical photosystems have been designed to reduce oxygen to hydrogen peroxide. Although hydrogen peroxide finds its main applications as a rather expensive bleaching agent, it can also be considered as an energy-rich compound since it liberates more than 100 kJ/mol when it decomposes into water and oxygen.

In biological photosystems (whole cells, thylakoids or isolated PSI) a simple redox catalyst transfers electrons from the terminal acceptor of PSI to oxygen, the light-transducing system being the natural photosynthetic apparatus itself. In artificial model systems, by contrast, it is a redox photocatalyst (flavin or Ru(II)-tris(2,2'-bipyridine)) who promotes the light-driven transfer of electrons from appropriate electron donors to molecular oxygen.

Laser flash photolysis has been used to study the primary photochemical reactions involving the excited state of the photosensitizers as well as the photochemically generated intermediate species. A sequence of reactions leading to oxygen photoreduction, with the concomitant formation of hydrogen peroxide, is proposed in every case.

INTRODUCTION

The best "sunlight-converting machine" that we know of is the chloroplast of green plants, but its overall efficiency at converting solar energy into stable chemical energy is well below 1% and so there is certainly room for new approaches where artificial photochemical techniques may have their chance. The development of product-oriented biological and chemical photosynthetic systems has actually aroused serious interest over the last few years (Connolly, 1981; Grätzel, 1983; Hall et al., 1983).

The main light-driven redox process of interest is water--splitting for hydrogen production, but the production of other energy-rich compounds such as hydrogen peroxide also deserves attention (Claesson and Hölmstrom, 1982; Grätzel, 1983). Hydrogen peroxide is not only an useful source of oxygen widely used in industry but also a powerful source of energy capable of releasing more than 100 kJ/mol upon decomposition into water and oxygen; in fact, highly concentrated peroxide solutions (85% or even higher) have been used for rocket propulsion (Crampton et al., 1977).

In this report we describe a number of biological and chemical photosynthetic systems to produce hydrogen peroxide. The biological systems are based on the use of the natural photosynthetic apparatus to promote the endergonic reduction of oxygen to hydrogen peroxide with electrons from water (the well-known Mehler reaction); the process is accelerated by several redox catalysts such as flavins, viologens and others. In contrast, the artificial model systems are based on the use of a visible light-absorbing compound (flavin or Ru(II)-tris(2,2'-bipyridine)) as the photosensitizer driving the transfer of electrons from apppropriate electron donors to molecular oxygen.

MATERIALS AND METHODS

H_2O_2 photoproduction by whole cells was determined by using <u>Anacystis nidulans</u> strain L-1402-1 (Göttingen University) grown on the synthetic medium of Guerrero et al. (1974). For experiments, harvested cell suspensions were centrifuged and resuspended in a medium C-m containing only the major salts. The H_2O_2-forming reaction was carried out in a 1-cm optical path spectrofluorometric cuvette containing 1 ml of cell suspension (8 μg chlorophyll) and 2 ml of 50 mM phosphate buffer, pH 7,5, supplemented with 5 units of peroxidase and 6 mg of scopoletin. Methyl viologen at 0.1 mM final concentration was used as the redox catalyst. Peroxide formation was determined by following the decrease in the scopoletin fluorescence at 460 nm upon irradiation of the reaction cell with orange light according to Patterson and Myers (1973).

H_2O_2 formation by tylakoids or isolated PSI was followed in a transparent lucite cuvette as previously described

(De la Rosa et al., 1986; Navarro et al., 1987a). Riboflavin at 10 μM final concentration was used as the redox catalyst. Thylakoid and PSI suspensions were prepared from spinach bought in the market following the procedures of Arnon and Chain (1977) and Peters et al. (1983), respectively.

For H_2O_2 photoproduction by artificial model systems, 0.1 mM flavin or Ru(II)-tris(2,2'-bipirydine) was used as the photosensitizer and 20 mM semicarbazide as the electron donor. All solutions were in 1 M NaOH. The photochemical reaction was followed as previously described (Navarro et al., 1987b; 1987c).

347-nm laser flash photolysis experiments were performed by means of a pulsed ruby laser (Navarro et al., 1987b; 1987c).

RESULTS AND DISCUSSION

The light-promoted endergonic reduction of oxygen to hydrogen peroxide with electrons from water as driven by the biological photosynthetic apparatus (whole cells or thylakoids) was first considered to be a photosystem that, for its simplicity and cleanness, could be most suitable for the conversion and storage of solar energy (see Fig. 1). The experimental data obtained when measuring the initial rates of hydrogen peroxide formation by such biological systems in the

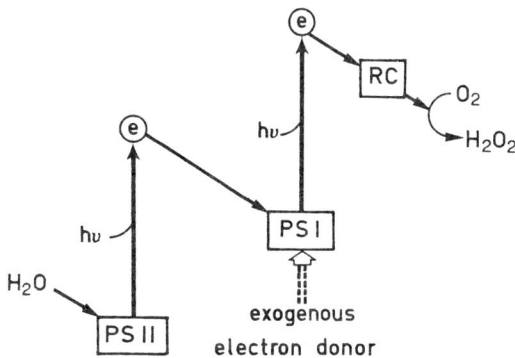

Fig. 1 Schematic drawing of hydrogen peroxide photoproduction by the biological photosynthetic apparatus with electrons either from water or from an exogenous electron donor. A redox catalyst (RC) transfers electrons from the terminal acceptor of photosystem I to molecular oxygen.

presence of methyl viologen (whole cells) or riboflavin (thylakoids) as the redox catalyst are shown in Table 1.

TABLE 1 Initial rates of hydrogen peroxide photoproduction by biological and chemical systems. For experimental conditions, see Materials and Methods.

Photosystem	Electron Donor	H_2O_2 production*
Biological		
whole cells	water	58
thylakoids	water	15
thylakoids	DTE/TMPD	648
isolated PSI	DTE/TMPD	1836
Chemical		
lumiflavin	semicarbazide	222
$Ru(bpy)_3^{2+}$	semicarbazide	167

*Initial rates of hydrogen peroxide production by biological and chemical photosystems are expressed in μmol mg(chl)$^{-1}$ h^{-1} and μmol l^{-1} min^{-1} H_2O_2 formed, respectively.

In all cases H_2O_2 formation ceases in a short time (no more than 10 minutes) because of photoinhibition of the photosynthetic electron transport activity, which is more pronounced at photosystem II than at photosystem I (De la Rosa et al., 1986). The better stability of PSI as compared to that of PSII was then taken advantage of in devising hydrogen peroxide production systems using thylakoid or isolated PSI suspensions with an electron source other than water (see also Fig. 1). As shown in Table 1, the initial rates of peroxide formation by thylakoids or photosystem I with DTE-reduced TMPD as the electron-donating system are significantly increased with respect to those obtained with water as the electron source; in addition, H_2O_2 photoproduction is quite slower to reach a stable plateau (data not shown).

As an alternative to biological systems, several artificial photosystems were designed to produce hydrogen peroxide with higher simplicity and efficiency. Among different

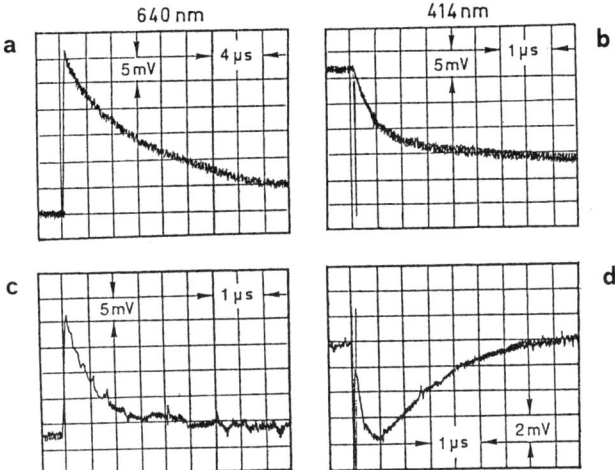

Fig. 2 Oscilloscope traces showing lumiflavin (LF) triplet decay at 640 nm (left) and flavosemiquinone formation at 414 nm (right). a: LF alone; b,c: LF + semicarbazide; d: LF + oxygen + semicarbazide. The concentrations used were: LF, 25 μM; oxygen, 1.26 mM; semicarbazide, 20 mM. All solutions were in 1 M NaOH.

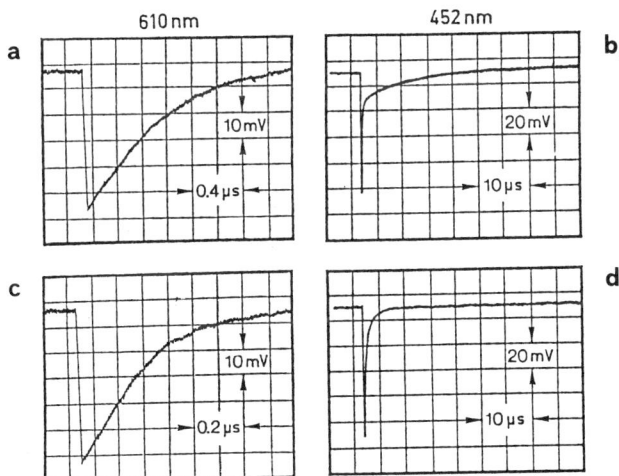

Fig. 3 Oscilloscope traces showing emission of light at 610 nm by the photoexcited ruthenium complex (left) and its bleaching recovery at 452 nm (right). a,b: Ru(II)-tris(2,2'-bipyridine) alone; c: a + oxygen; d: a + semicarbazide. The concentrations used were: Ru(II)--tris(2,2'-bipyridine), 50 μM; oxygen, 1.26 mM; semicarbazide, 20 mM. All solutions were in 1 M NaOH.

types of visible light-absorbing compounds assayed as photosensitizers (Navarro et al. 1987a), lumiflavin and Ru(II)-tris(2-2'-bipyridine) were shown to be the most efficient ones. Table 1 shows the initial rates of H_2O_2 formation by such photosensitizers with semicarbazide as a sacrificial electron donor.

A profound understanding of the reaction mechanism through which H_2O_2 is being formed could obviously enable optimization and lead to a maximal efficiency of such photochemical systems. Laser flash photolysis was thus employed to study the primary and secondary photochemical reactions in which lumiflavin and Ru(II)-tris(2,2'-bipyridine) are involved. Fig. 2 shows that photoexcited lumiflavin in its triplet state (picture a) is efficiently quenched by semicarbazide (picture c); this leads to flavosemiquinone formation (picture b). Further addition of molecular oxygen allows oxidation of flavosemiquinone radicals (picture d). On the other hand, Fig. 3 shows that photoexcited Ru(II)-tris(2,2'-bipyridine) is efficiently quenched by molecular oxygen (picture c) and the so

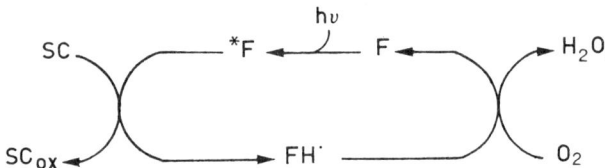

Reductive-quenching

Oxidative-quenching

Fig. 4 Proposed mechanism of hydrogen peroxide formation by the photochemical systems. While lumiflavin appears to react via reductive quenching by semicarbazide, the ruthenium complex appears to react via oxidative quenching by molecular oxygen.

formed oxidized ruthenium complex is quickly reduced by semicarbazide (picture d).

On the basis of these data one can clearly state that lumiflavin is submitted to a reductive quenching by semicarbazide, while Ru(II)-tris(2,2'-bipyridine) reacts via oxidative quenching by molecular oxygen (Navarro et al., 1987b; 1987c). Fig. 4 shows a simplified sequence of reactions which leads to oxygen photoreduction, with the concomitant formation of hydrogen peroxide, with electrons from semicarbazide as driven by lumiflavin and Ru(II)-tris(2,2'-bipyridine). As can be seen in the figure, the same net result is obtained independently of the type of quenching to which the photosensitizer is submitted.

Acknowledgements

This work has been supported by INTEROX química S.A. (c/o. SOLVAY & Cie, Belgium).

REFERENCES

Arnon, D.I. and Chain, R.K. 1977. Ferredoxin-catalyzed photophosphorylations: concurrence, stoichiometry, regulation and quantum efficiency. In "Photosynthetic Organelles" (Ed. S. Miyachi, S. Katoh, Y. Fujita and K. Shibata). (Japanese Society of Plant Physiologists, Japan). p. 433.

Claesson, S. and Hölmstrom, B. (Eds.) 1982. solar energy--photochemical processes available for energy conversion. (National Swedish Board for Energy Source Development, Uppsala).

Connolly, J.S. (Ed.) 1981. Photochemical conversion and storage of solar energy. (Academic Press, New York).

Crampton, C.A., Faber, G., Jones, R., Leaver, J.P. and Schelle, S. 1977. The manufacture, properties and uses of hydrogen peroxide and other inorganic peroxy compounds. In "The Modern Inorganic Chemical Industry" (Ed. R. Thompson). (The Chemical Society, London). pp. 232-272.

De la Rosa, M.A., Rao, K.K. and Hall, D.O. 1986. Hydrogen peroxide photoproduction by free and immobilized spinach thylakoids. Photobiochem. Photobiophys., 11, 173-187.

Grätzel, M. (Ed.) 1983. Energy resources through photochemistry and catalysis. (Academic Press, New York).

Guerrero, M.G., Manzano, C. and Losada, M. 1974. Nitrite photoreduction by a cell-free preparation of Anacystis nidulans. Plant Sci. Lett., 3, 273-278.

Hall, D.O., Palz, W. and Pirrwitz, D. (Eds.) 1983. Photochemical, photoelectrochemical and photobiological processes. Series D. Vol. 2. (Reidel, Dordrecht).

Navarro, J.A., Roncel, M., De la Rosa, F.F. and De la Rosa, M.A. 1987a. Light-driven hydrogen peroxide production as a way to solar energy conversion. Bioelectrochem. Bioenerg., in press.

Navarro, J.A., Roncel, M., De la Rosa, F.F. and De la Rosa, M.A. 1987b. Hydrogen peroxide photoproduction by the semicarbazide-tris(2,2'-bipyridine)ruthenium(II)-oxygen system. J. Photochem. Photobiol., in press.

Navarro, J.A., Roncel, M. and De la Rosa, M.A. 1987c. Potentiometric and laser flash photolysis studies of the pH dependence of hydrogen peroxide production by the semicarbazide-lumiflavin-oxygen photosystem. Photochem. Photobiol., in press.

Patterson, C.O.P. and Myers, J. 1973. Photosynthetic production of hydrogen peroxide by Anacystis nidulans. Plant Physiol., 51, 104-109.

Peters, A.L.J., Wielink, J.E.V., Sang, H.W.W.F., De Vries, S. and Kraayenhof, R. 1983. Studies on well coupled photosystem I-enriched subchloroplast vesicles; content and redox properties of electron-transfer components. Biochim. Biophys. Acta, 722, 460-470.

CURRENT TOPICS IN THE BIOCHEMISTRY OF Fe-HYDROGENASES

W.R. Hagen

Department of Biochemistry, Agricultural University

De Dreijen 11, NL-6703 BC Wageningen, The Netherlands

ABSTRACT

The bacterial activation of protons/molecular hydrogen is a process of potential interest for the practical production of energy-rich compounds. Our current knowledge on the key enzyme in this process is diffuse reflecting differences in molecular properties of hydrogenases from different sources. Non-nickel hydrogenases form a small, relatively homogeneous subgroup. Desulfovibrio vulgaris (Hildenborough) hydrogenase is of special interest because it exhibits the highest hydrogen-production activity and can be isolated in a form inert towards atmospheric oxygen. Recent studies on the stability, the mechanism of action and the biosynthesis of this and related proteins are discussed.

INTRODUCTION

Following the 1973 global oil crisis fundamental research efforts towards an understanding of the biological mechanism of molecular-hydrogen activation have increased exponentially. Remarkably, although this extensive exploration has provided us with a wealth of details on more than a handful of hydrogenases, at present we still appear to be far away from a unifying grasp of how the enzyme operates. The last comprehensive review on the subject (Adams et al., 1981) has reached now the respectable age of over six years (it makes no mention of nickel) and no attempt to a follow up is in progress. Our desire to be inspired by the bacterial H2 chemistry in finding useful contributions to the resolution of our ongoing problem of energy requirement is frustrated by this lack of a unifying descriptive framework. We stealthily ask ourselves the question whether there is such a thing as "THE" hydrogenase. The enzyme H2:acceptor oxidoreductase, when isolated from different species, or even from different strains, displays a discouraging diversity in its enzymic, molecular, and spectroscopic properties.

However, the soluble hydrogenases from the sulfate-reducing anaerobe Desulfovibrio vulgaris, strain Hildenborough, and from the fermentative anaerobes Megasphaera elsdenii and Clostridium pasteurianum appear to form a small subgroup of closely related entities. All three have been reported to contain about twelve Fe and about twelve acid-labile sulfide ions and to lack Ni (and probably selenium). Therefore, they are named Fe-hydrogenases, or "Fe-only" hydrogenases, to set them apart from the Ni- or NiFe-hydrogenases. They also share a relatively high (in comparison with other hydrogenases) H2 production activity; therefore they are named bidirectional hydrogenases, in contrast with the so called uptake hydrogenases. Thus, to all reason the combined research efforts on these three proteins are to provide uniform information on THE mechanism of bidirectional hydrogenases, or more specifically, on the properties of THE site (presumably an Fe/S cluster; see below) of hydrogen activation.

Three more hydrogenases have been reported to contain Fe and to lack Ni, namely, from D. vulgaris, strain Myazaki (Yagi et al., 1985), from Acetobacter woodii (Ragsdale and Ljungdahl, 1984), and hydrogenase II from C. pasteurianum (Adams and Mortenson, 1984). The relation of these proteins to the subgroup of bidirectional hydrogenases is at present not clear.

Within the three-membered subgroup the bidirectional hydrogenase from D. vulgaris (Hildenborough) is possibly the more interesting one from a biotechnological point of view because it exhibits the highest H2 production activity and because it can be isolated in a resting, O2 stable form. Below we give a status report on our ongoing attempts to understand the complex biochemistry of this enzyme with emphasis on phenomena that may bear relevance to future practical applications.

STRUCTURE OF THE HYDROGEN ACTIVATING SITE

A knowledge of the structure of the H2-activating site is not only important for an understanding of the mechanism of action of the enzyme but is relevant for practical applications since it may provide clues on how to improve on, e.g., O2- and pH-instability. From a different perspective the active site may be a model to be mimicked by the inorganic chemist thus altogether eliminating the biological aspects of the problem.

At present there is no detailed knowledge available on the active site of any hydrogenase. Although the bidirectional hydrogenases are generally considered to contain no other metal than Fe it is not known how many Fe are present in the active site. What we do know is the amino-acid sequence for D. vulgaris hydrogenase (Voordouw and Brenner,

1985) not for the other two enzymes. An N-terminal stretch containing eight Cys residues shows striking sequence homology with 2[4Fe-4S] bacterial ferredoxins. Combined with similarity in EPR spectra this strongly suggests the existence of two ferredoxin cubanes in the enzyme. The total number of Fe/S has been thought for some years to be 12 (i.e. a multiple of 4), however, extensive determinations of Fe, S2-, protein and molecular mass have recently led us to correct this number to approximately 14 (Hagen et al., 1986; Filipiak et al., 1987) 8 of which are in the two cubanes. For two reasons we think that the remaining 5 to 7 Fe form the active site. Firstly, when the two structural genes for D. vulgaris hydrogenase are cloned and expressed in Escherichia coli the purified product appears to be identical to the original hydrogenase except that activity is missing and the EPR signals (see below) other than the two-cubane signal are missing (Voordouw et al., 1987). Secondly, the activity of M. elsdenii hydrogenase is approximately linearly related to its total-minus-eight Fe content (Filipiak et al. 1987). In all these preparations the two-cubane EPR signal had its full intensity equivalent to two $S=1/2$ systems (Hagen et al., unpublished).

There is no known 5-7 Fe containing structure in Fe/S proteins. A series of 6 Fe/S model compounds have been synthesized, however, their paramagnetic properties appear to be quite different (Kanatzidis et al., 1985) from those of hydrogenase (Hagen et al., 1986). The active-site structure is likely to come from X-ray analysis of crystallized hydrogenase. The first successful crystallization of an hydrogenase was recently reported for the enzyme from D. vulgaris, strain Myazaki (Higuchi et al., 1987). No amino acid sequence data are available for this protein. Over the last two years we have repeatedly produced crystals of resting hydrogenase from D. vulgaris, strain Hildenborough, however, all crystals showed internal disorder sufficient to preclude X-ray analysis (collaboration with W.G. Hol et al.). This protein was homogeneous according to the FPLC test used for the Myazaki protein. Attempts to find conditions that produce ordered crystals will continue.

STABILITY VERSUS MOLECULAR OXYGEN

In any application of biological hydrogen conversion one probably has to deal with unwanted side reactions with molecular oxygen. Fe-hydrogenase is isolated from D. vulgaris in an O2-stable form by washing cells with Tris/EDTA pH 9 at 30 C (van der Westen et al., 1978). This "resting" enzyme is activated by reduction with H2 wereupon it becomes just as extremely O2-sensitive as the other two bidirectional hydrogenases. O2-insesitivity is regainable by anaerobic reoxidation with 2,6-dichlorophenol indophenol in the presence of iron and EDTA followed by addition of Tris/EDTA pH 9 (van Dijk et al., 1983). The mechanism of this activation/deactivation is not understood at all.

The three Fe/S domains would be natural candidates to inspect for structural changes related to the reactivity towards molecular oxygen. Fe/S clusters can be studied by magnetic-resonance spectroscopies at relatively high concentration > 10 µM. Unfortunately, we have found the deactivation process of D. vulgaris hydrogenase to be reproducible only at enzyme concentrations of < 10 µM. Therefore, we have - on a molecular level - only been able to study the mechanism of activation, not of deactivation.

In the resting enzyme all iron is in diamagnetic environments as shown by quantitative, high-resolution Mossbauer spectroscopy from 4 to 175 K (W.R. Dunham et al., unpublished). In reduction and reoxidation the two [4Fe-4S] clusters cycle between their common $S=1/2$; $S=0$ spin states. However, the remainder of the Fe (the putative active site) ends up in the activated enzyme in a paramagnetic state with complex EPR spectrum. Most of this Fe gives a signal that we tentatively attribute to an $S=2$ state (Hagen et al., 1986). This assignment would imply that O2-sensitivity is obtained parallel with a spin-state transition of the active center from $S=0$ to $S=2$. Spin state transitions usually indicate the occurrence of a significant conformational change. It would seem reasonable to assume that deactivation is the reverse of this conformational change. This hypothesis suggests that the (de)activation process can be studied by techniques that look at the protein rather than at the iron and studies along these lines are in progress.

Our present working hypothesis of a single-step conformational change is probably an oversimplification of reality. When we reoxidize D. vulgaris hydrogenase with solvent protons, by replacing the reductant H2 with argon, we observe in addition to the $S=2$ EPR signals three more EPR spectral species of substoichiometric intensity: rhombic-1 ($g=$ 2.11; 2.05; 2.00), axial ($g=$ 2.06; 2.01) and rhombic-2 ($g=$ 2.07; 1.96; 1.89). The axial species is of very low intensity but slowly increases during repeated redox cycling over several hours. We therefore think that this signal represents a breakdown product. This interpretation is at variance with the work of others who have claimed the signal to represent a physiologically relevant species (Huynh et al., 1984; Patil et al., 1986). The rhombic-1 species is very similar to a signal observed for the oxidized form of the other two bidirectional hydrogenases, however, its intensity in the D. vulgaris enzyme is lower by an order of magnitude. The rhombic-2 signal has not hitherto been reported; it is more difficult to detect than the other two signals because of its larger line width and g anisotropy.

In summarizing this section it appears that the activation/O2-sensitization and, possibly, also its reversal be ameanable now to detailed systematic studies on the molecular level. The minimum hypothesis of a single conformational event will likely have to be extended to include several intermediates.

Scheme I
Redox stages in the activation of D.vulgaris hydrogenase

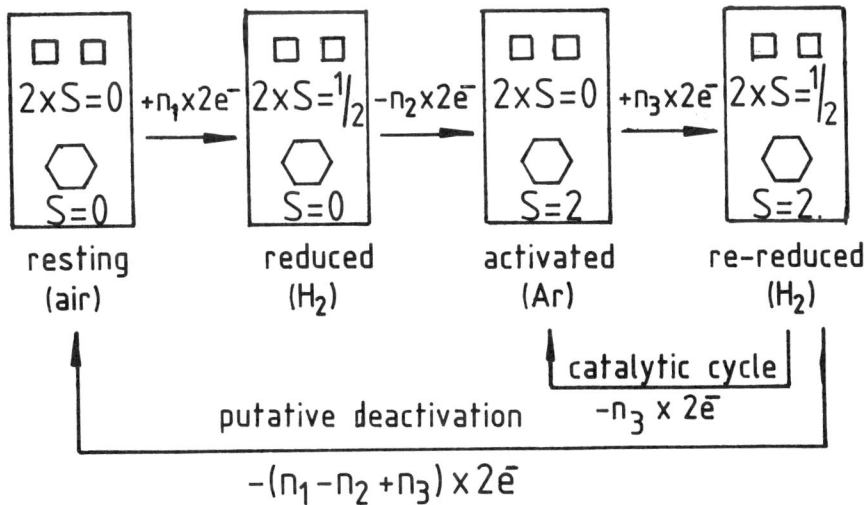

ELECTRON TRANSFER PATHWAYS

Why does an enzyme need three redox groups to catalyze a reaction involving only two electrons? The simplest explanation would be that the two cubanes are just the electronic coupling between the H2-producing site and the source of reducing equivalents. This implies that the third cluster can transfer two electrons, a property not previously observed for Fe/S clusters in proteins. The pathway(s) of electrons through any hydrogenase is unknown. Also, the number of reducing equivalents that can be stored in hydrogenase is unknown. For D. vulgaris hydrogenase at least four states can be discerned with EPR spectroscopy after equilibration with different gas phases as shown in Scheme I (□ is a ferredoxin cubane; ◇ symbolizes the active site of 5-7 Fe). All depicted steps involve transfer of two electrons to or from the pair of ferredoxin cubanes. However, the transfer of electrons to/from the active site cluster may be either 0 or 2 (the unknowns n1, n2, n3 are each either 1 or 2). After a single reduction/oxidation cycle the S=2 EPR signal appears and does not change upon subsequent incubation under H2. Thus the redox state of the active-site cluster in any of these four states is undetermined. It appears that the transfer of electrons from the ferredoxin cubanes to protons through the active site is not observable in these slow-equilibration experiments. Note that this interpretation holds under the model that the Fe in the active site forms a single, uniform entity, i.e. the minor EPR species mentioned above are not (yet) included in the scheme.

A potentially promising route to learn more about the complex redox chemistry of hydrogenase is to look at the direct electron transfer between the enzyme and an electrode. A study of this phenomenon may also be relevant to a possible coupling of solar cells with hydrogenase-catalyzed H2 production. Results from this line of research have thus far been very limited. A response was obtained in differential-pulse polarography on the dropping mercury electrode modified with polylysine (cf. van Dijk et al., 1985). Attempts to use more biologically-friendly electrodes remained unsuccessful over the last several years. To understand the reasons for this failure the electron transfer between D. vulgaris hydrogenase and viologens of different charge was recently studied by means of cyclic voltammetry and chronoamperometry using a glassy carbon working electrode. The results were interpreted in terms of an electrostatic interaction between the mediator and a negatively charged part of the protein (Hoogvliet et al., submitted). To overcome electrostatic repulsion between the electrode and the hydrogenase we continue our attempts to try to modify suitable electrodes and to look for significant responses. We have very recently obtained a direct, unmediated response on a modified edge-graphite electrode from the reduction of activated (H2/Ar-cycled) D. vulgaris hydrogenase (I.S.M. Psalti et al., unpublished) and we will explore this effect in more detail.

THE BIOSYNTHESIS OF HYDROGENASE HOLOENZYME

Anaerobes are not particularly collaborative coworkers when one aims at ultimately applying a biological system for practical purposes. The genes for D. vulgaris hydrogenase were cloned in E. coli (Voordouw et al., 1985) initially for fundamental molecular biological studies. However, it is also our explicit purpose to produce more hydrogenase more easily and more cheaply. In the short run this would facilitate fundamental experiments requiring larger amounts of protein such as crystallization experiments and certain spectroscopic studies, e.g., EXAFS spectroscopy. In the long run the production of bidirectional hydrogenase by, e.g., E. coli may be a conditio sine qua non to guarantee a significant biotechnological role for the enzyme. Two major problems have frustrated thus far our attempts to "go aerobic".

The first problem has already been mentioned above: the gene product isolated and purified from E. coli appears to be identical to the original enzyme except for the lack of the active site Fe/S cluster (Voordouw et al., 1987). Apparently, E. coli does have the biosynthetical machinery to construct the two ferredoxin cubanes in their proper conformation, however, it is unable to provide the chemical and/or genetic conditions to synthesize the special 5-7 Fe cluster. In analogy to the complex genetics of the enzyme nitrogenase we consider it likely that the information carried by the two structural genes for the D. vulgaris hydrogenase is insufficient to produce an active holoenzyme. In other words there is a requirement for one or more extra gene products that are involved in the synthesis and/or building-in of the special Fe/S cluster. Attempts to reconstitute the inactive protein in the test tube with Fe/S/reducing equivalents plus extracts from D. vulgaris cells have thus far been without succes (W.R. Hagen et al., unpublished). A program is underway to try to restore hydrogenase activity in a hydrogenase-negative mutant of E. coli by cloning D. vulgaris genes in addition to the structural genes for bidirectional hydrogenase (W.M.A.M. van Dongen et al., unpublished).

The second problem in the synthesis of active recombinant hydrogenase is the observation of an impaired transport over the periplasmatic membrane in E. coli. From blotting experiments of cell extracts and purified hydrogenase it was estimated that the ratio of hydrogenase protein per total protein should be an order of magnitude higher in the E. coli clone compared to D. vulgaris. Employing identical isolation procedures the yield per cell mass from the two species was found to be approximately the same. Cell fractionation experiments revealed the following picture: only a small portion of the two-subunit recombinant enzyme cofractionates with the periplasma; the majority of the larger subunit is detected in the cytoplasmic fraction; most of the smaller subunit is found in precursor form in the membrane fraction (van Dongen et al., unpublished). In E. coli clones in which the gene for either one of the subunits has been deleted, the export of the other subunit is completely abolished.

The hydrogenase larger subunit (46 kDa) does not carry a presequence (Voordouw and Brenner, 1985). The smaller subunit (10 kDa) is synthesized as a precursor (13.5 kDa) with a presequence of 34 amino-acid residues (Prickril et al., 1986). Bacterial export sequences generally consist of a short, charged head followed by a very hydrophobic stretch of some 20 residues, followed by a 6 residue stretch defining the cleavage site (cf. von Heine, 1985). We do not find this structure in the presequence of D. vulgaris hydrogenase. There is no clearly hydrophobic region of significant length. However, when we assume the presequence to be an alpha-helix then we find two groupings of positively charged residues on one side of the chain. The majority of hydrophobic residues is located on the opposite side of the chain (van Dongen et al., unpublished). This specific signal is reminiscent of the so called "amphiphilic" signal sequences as, e.g., found in targeting sequences for mitochondrial import proteins in eukaryotes (cf. Schatz, 1987). We hypothesize that the specificity of the hydrogenase signal sequence might be required for interaction with a specific receptor that might be involved in subunit association, membrane translocation, or building-in of the Fe/S active site cluster.

We continue our attempts to gain some control over the construction of D. vulgaris bidirectional hydrogenase in E. coli not only to come to an understanding of the biosynthesis of complex Fe/S enzymes, but also to bring the candidacy of this protein for biotechnological applications within the realm of possibility.

REFERENCES

Adams, M.W.W., Mortenson, L.E. and Chen, J.-S. (1981) Biochim. Biophys. Acta 594, 105-176
Adams, M.W.W. and Mortenson, L.E. (1984) J. Biol. Chem. 259, 7045-7055
Filipiak, M., Hagen, W.R., Grande, H.J., Dunham, W.R., van Berkel-Arts, A., Kruse-Wolters, K.M. and Veeger, C. (1987) Recl. Trav. Chim. Pays-Bas 106, 230
Hagen, W.R., van Berkel-Arts, A., Kruse-Wolters, K.M, Dunham, W.R. and Veeger, C. (1986) FEBS Lett. 201, 158-162
Hagen, W.R., van Berkel-Arts, A., Kruse-Wolters, K.M., Voordouw, G. and Veeger, C. (1986) FEBS Lett. 203, 509-514
Higuchi, Y., Yasuoka, N., Kakudo, M., Katsube, Y., Yagi, T. and Inokuchi, H. (1987) J. Biol. Chem. 262, 2823-2825
Huynh, B.H., Czechowski, M.H., Kruger, H.-J., DerVartanian, D.V., Peck, H.D. and LeGall, J. (1984) Proc. Natl. Acad. Sci. USA 81, 3728-3732
Kanatzidis, M.G., Hagen, W.R., Dunham, W.R., Lester, R.K. and Coucouvanis, D. (1985) J. Am. Chem. Soc. 107, 953-961
Patil, D.S., Czechowski, M.H., Huynh, B.H., LeGall, J., Peck, H.D. and DerVartanian, D.V. (1968) Biochem. Biophys. Res. Commun. 137, 1086-1093
Prickril, B.C., Czechowski, M.H., Przybyla, A.E., Peck, H.D. and LeGall, J. (1986) J. Bacteriol. 167, 722-725
Ragsdale, S.W. and Ljungdahl, L.G. (1984) Arch. Microbiol. 139, 361-365
Schatz, G. (1987) Eur. J. Biochem. 165, 1-6
Van der Westen, H.M., Mayhew, S.G. and Veeger, C. (1978) FEBS Lett. 86, 122-126
Van Dijk, C., van Berkel-Arts, A. and Veeger, C. (1983) FEBS Lett. 156, 340-344
Van Dijk, C., Laane, C. and Veeger, C. (1985) Recl. Trav. Chim. Pays-Bas 104, 245-252
Von Heine, G. (1984) J. Mol. Biol. 184, 99-105
Voordouw, G., Walker, J.E. and Brenner, S. (1985) Eur. J. Biochem. 148, 509-514
Voordouw, G. and Brenner, S. (1985) Eur. J. Biochem. 148, 515-520
Voordouw, G., Hagen, W.R., Kruse-Wolters, K.M., van Berkel-Arts, A. and Veeger, C. (1987) Eur. J. Biochem. 162, 31-36
Yagi, T., Kimura, K. and Inokuchi, H. (1985) J. Biochem. (Tokyo) 97, 181-187

List of Participants

Dr. H. Bottin
Service Biophysique
Ctr d'Etudes Nucleaires de Saclay
91191 Gif-Sur-Yvette Cedex
FRANCE

Prof. Dr. S. Costa
Centro di Quimico Estrutural
Instituto Superior Tecnico
Complexo I
P-1096 Lisboa Cedex
PORTUGAL

Dr. M.A. De la Rosa
Departamento de Bioquimica
Facultad de Biologia Y.C.S.I.C.
Universidad de Sevilla
Apartado 1095
41080 Sevilla
SPAIN

Dr. D. Dietrich-Buchecker
Lab. de Chimie Organo-Minerale
Institut de Chimie
Universite Louis Pasteur
1, rue Blaise Pascal
67008 Strasbourg Cedex
FRANCE

Prof. M.C.W. Evans
Department of Botany & Microbiology
University College London
Gower St.
London WC1 6BT
UK

Dr. A. Fernandez
Instituto de Ciencias de Materiales
Centro Mixto
Univ. Sevilla - C.S.I.C.
Apartado 1115
41071 Sevilla
SPAIN

Dr. A. Galindo
Departamento de Chimica Inorganica
Facultad de Quimica
Universidad de Sevilla
Apartado 553
Sevilla
SPAIN

Dr. L. Giorgi
The Royal Institution
21 Albemarle Street
London W1X 4BS
UK

Prof. M. Gratzel
Department of Chemistry
Institute of Physical Chemistry
EPFL - Ecublens
CH - 1015 Lausanne
Switzerland

Prof. M. Guerrero
Department of Biochemistry
Universidad de Sevilla
Apartado 1095
Sevilla
SPAIN

Dr. W.R. Hagen
Department of Biochemistry
Agricultural University
De Dreijen 11
NL 6703 BC Wageningen
THE NETHERLANDS

Prof. D.O. Hall
King's College London
University of London
Campden Hill Road
London W8 7AH
UK

Dr. L. Häussling
Institute for Organic Chemistry
Universitat Mainz
D-6500 Mainz
F.R. GERMANY

Dr. W. Jaegermann
Hahn-Meitner Institute
Bereich Strahlen Chemie
Glienicker Strasse 100
Postfach 390128
D-1000 Berlin 39
F.R. GERMANY

Dr. A. Juris
c/o Prof. V. Balzani
Universita Degli Studi di Bologna
Dipartimento di Chimica G Ciamican
via Selni, 2
40126 Bologna
ITALY

Dr. J.M. Kelly
Department of Chemistry
Trinity College Dublin
Dublin 2
IRELAND

Dr. S. Lingier
Rijks Universiteit Gent
Fysische Scheikunde S2
Krijgslaan 271
B-9000 Gent
BELGIUM

Prof M. Losada
Department of Biology
Faculty of Biology
Universidad de Sevilla
Apartado 1095
Sevilla
SPAIN

Prof. A. Mackor
TNO
Institute of Applied Chemistry
P.O. Box 5009
NL-3502 JA Utrecht
THE NETHERLANDS

Prof. P. Mathis
Service Biophysique
Centre d'Etudes Nucleaires de Saclay
91191 Gif-Sur-Yvette Cedex
FRANCE

Dr. D. Meissner
Institute for Physical Chemistry
  of the University of Hamburg
Bundesstrasse 45
D-2000 Hamburg 13
F.R. GERMANY

Dr. J.A. Navio-Santos
Departamento de Quimica General
Facultad de Quimica
Universidad de Sevilla
Sevilla
SPAIN

Prof. Sir George Porter
The Royal Society
6 Carlton House Terrace
London SW1
UK

Dr. J.M. Ramirez
Centro de Investigaciones Biologicas
Velazquez 144
28006 Madrid
SPAIN

Dr. K.K. Rao
Dept of Biology
King's College London
Campden Hill Road
London W8 7AH
UK

Prof. P. Salvador
Consejo Superior de Investigaciones
  Cientificas
Inst. de Catalysis y Petroleoquimica
Serrano 119
28006 Madrid
SPAIN

Dr. F. Scandola
Dipartimento di Chimica
Universita di Ferrara
Via L. Borsari 46
44100 Ferrara
ITALY

Dr. J.L. Serra
Departmento di Bioquimica
Facultad de Ciencias
Universidad del Pais Vasco
Apartado 644
Bilbao
SPAIN

Dr. D. Simpson
Department of Physiology
Carlsberg Laboratory
Gamle Carlsberg Vej 10
DK-2500 Copenhagen Valby
DENMARK

Prof J.W. Verhoeven
Laboratory of Organic Chemistry
University of Amsterdam
Nieuwe Achtergracht 129
NL-1018 WS Amsterdam
THE NETHERLANDS

Dr. E. Vrachnou
Nuclear Research Centre Demokritos
Department of Chemistry
Aghia Paraskevi
Athens
GREECE

Dr. F.A. Wollman
Inst. de Biologie Physico-Chimique
Fondation Edmond de Rothschild
13, Rue Pierre et Marie Curie
75005 Paris
FRANCE

RAYMOND H. FOGLER LIBRARY